T0344905

Meta-Heuristic and Evolutionary Algorithms for Engineering Optimization

Wiley Series in Operations Research and Management Science

A complete list of the titles in this series appears at the end of this volume

Meta-Heuristic and Evolutionary Algorithms for Engineering Optimization

Omid Bozorg-Haddad
University of Tehran, Iran

Mohammad Solgi
University of Tehran, Iran

Hugo A. Loáiciga
University of California, Santa Barbara, USA

This edition first published 2017

Registered Office
John Wiley & Sons, Inc., 111 River Street, Hoboken, NJ 07030, USA

Editorial Office
111 River Street, Hoboken, NJ 07030, USA

For details of our global editorial offices, customer services, and more information about Wiley products visit us at www.wiley.com.

Wiley also publishes its books in a variety of electronic formats and by print-on-demand. Some content that appears in standard print versions of this book may not be available in other formats.

Library of Congress Cataloguing-in-Publication Data

Names: Bozorg-Haddad, Omid, 1974– author. | Solgi, Mohammad, 1989– author. | Loaiciga, Hugo A., author.
Title: Meta-heuristic and evolutionary algorithms for engineering optimization / Omid Bozorg-Haddad, Mohammad Solgi, Hugo A. Loáiciga.
Description: Hoboken, NJ : John Wiley & Sons, 2017. | Series: Wiley series in operations research and management science | Includes bibliographical references and index. |
Identifiers: LCCN 2017017765 (print) | LCCN 2017026412 (ebook) | ISBN 9781119387077 (pdf) | ISBN 9781119387060 (epub) | ISBN 9781119386995 (cloth)
Subjects: LCSH: Mathematical optimization. | Engineering design–Mathematics.
Classification: LCC QA402.5 (ebook) | LCC QA402.5 .B695 2017 (print) | DDC 620/.0042015196–dc23
LC record available at https://lccn.loc.gov/2017017765

Cover Design: Wiley
Cover Images: (Top Image) © Georgina198/Gettyimages; (Bottom Image) © RomanOkopny/Gettyimagess

Set in 10/12pt Warnock by SPi Global, Pondicherry, India

Printed in the United States of America

10 9 8 7 6 5 4 3 2 1

Contents

Preface

Engineers search for designs of new systems that perform optimally and are cost effective or for the optimal operation and rehabilitation of existing systems. It turns out that design and operation usually involve the calibration of models that describe physical systems. The tasks of design, operation, and model calibration can be approached systematically by the application of optimization. Optimization is defined as the selection of the best elements or actions from a set of feasible alternatives. More precisely, optimization consists of finding the set of variables that produces the best values of objective functions in which the feasible domain of the variables is restricted by constraints.

Meta-heuristic and evolutionary algorithms, many of which are inspired by natural systems, are optimization methods commonly employed to calculate good approximate solutions to optimization problems that are difficult or impossible to solve with other optimization techniques such as linear programming, nonlinear programming, integer programming, and dynamic programming. Meta-heuristic and evolutionary algorithms are problem-independent methods of wide applicability that have been proven effective in solving a wide range of real-world and complex engineering problems. Meta-heuristic and evolutionary algorithms have become popular methods for solving real-world and complex engineering optimization problems.

Yet, in spite of meta-heuristic and evolutionary algorithms' frequent application, there is not at present a reference that presents and explains them in a clear, systematic, and comprehensive manner. There are several bibliographical sources dealing with engineering optimization and the application of meta-heuristic and evolutionary algorithms. However, their focus is largely on the results of application of these algorithms and less on their basic concepts on which they are founded. In view of this, it appears that a comprehensive, unified, and insightful overview of these algorithms is timely and would be welcome by those who seek to learn the principles and ways to apply meta-heuristic and evolutionary algorithms.

This book fills the cited gap by presenting the best-known meta-heuristic and evolutionary algorithms, those whose performance has been tested in many engineering domains. Chapter 1 provides an overview of optimization and illustration of its application to engineering problems in various specialties. Chapter 2 presents an introduction to meta-heuristic and evolutionary algorithms and their relation to engineering problems. Chapters 3–22 are dedicated to pattern search (PS), genetic algorithm (GA), simulated annealing (SA), tabu search (TS), ant colony optimization (ACO), particle swarm optimization (PSO), differential evolution (DE), harmony search (HS), shuffled frog-leaping algorithm (SFLA), honey-bee mating optimization (HBMO), invasive weed optimization (IWO), central force optimization (CFO), biogeography-based optimization (BBO), firefly algorithm (FA), gravity search algorithm (GSA), bat algorithm (BA), plant propagation algorithm (PPA), water cycle algorithm (WCA), symbiotic organisms search (SOS), and comprehensive evolutionary algorithm (CEA), respectively. The order of the chapters corresponds to the order of chronological appearance of the various algorithms, with the most recent ones receiving the larger chapter numbers. Each chapter describes a specific algorithm and starts with a brief literature review of its development and subsequent modification since the time of inception. This is followed by the presentation of the basic concept on which the algorithm is based and the mathematical statement of the algorithm. The workings of the algorithm are subsequently described in detail. Each chapter closes with a pseudocode of the algorithm that constitutes an insightful and sufficient guideline for coding the algorithm to solve specific applications.

Several of the algorithms reviewed in this book were developed decades ago, and some have experienced modifications and hybridization with other algorithms. This presentation is concerned primarily with the original version of each algorithm, yet it provides references that are concerned with modifications to the algorithms.

This book was written for graduate students, researchers, educators, and practitioners with interests in the field of engineering optimization. The format and contents chosen are intended to satisfy the needs of beginners and experts seeking a unifying, complete, and clear presentation of meta-heuristic and evolutionary algorithms.

Omid Bozorg-Haddad
Mohammad Solgi
Hugo A. Loáiciga

About the Authors

Omid Bozorg-Haddad
He is a professor at the Department of Irrigation and Reclamation Engineering of the University of Tehran, Iran (E-mail: OBHaddad@ut.ac.ir). His teaching and research interests include water resources and environmental systems analysis, planning, and management as well as application of optimization algorithms in water-related systems. He has published more than 200 articles in peer-reviewed journals and 100 papers in conference proceedings. He has also supervised more than 50 M.Sc. and Ph.D. students.

Mohammad Solgi
He received his M.Sc. degree in hydrological engineering from the University of Tehran, Iran (E-mail: Solgi_Mohammad@ut.ac.ir). His research interests include development of mathematical and computational techniques and their application in hydrology. He has published several articles in peer-reviewed journals and received awards for his works.

Hugo A. Loáiciga

He is a professor in the Department of Geography, University of California (Santa Barbara), United States (E-mail: Hugo.Loaiciga@ucsb.edu). His teaching and research interests include hydrology and water resources systems. He has published more than 250 articles in peer-reviewed journals and received numerous awards for his work.

List of Figures

Meta-heuristic and evolutionary algorithms are problem-independent optimization techniques. They are effective in solving a wide range of real-world and complex engineering problems. This book presents and explains the most important meta-heuristic and evolutionary algorithms known to date in a clear, systematic, and comprehensive manner. The algorithms presented in this book are pattern search (PS), genetic algorithm (GA), simulated annealing (SA), tabu search (TS), ant colony optimization (ACO), particle swarm optimization (PSO), differential evolution (DE), harmony search (HS), shuffled frog-leaping algorithm (SFLA), honey-bee mating optimization (HBMO), invasive weed optimization (IWO), central force optimization (CFO), biogeography-based optimization (BBO), firefly algorithm (FA), gravity search algorithm (GSA), bat algorithm (BA), plant propagation algorithm (PPA), water cycle algorithm (WCA), symbiotic organisms search (SOS), and comprehensive evolutionary algorithm (CEA). These algorithms are presented in a consistent and systematic format, explaining their applications to engineering optimization problems. This book provides students, researchers, and teachers with a comprehensive exposition of meta-heuristic and evolutionary algorithms with sufficient detail to understand their principles and apply them to specific problems.

Keywords:
Optimization
Engineering optimization
Meta-heuristic search
Evolutionary algorithms
Nature-inspired optimization algorithms

1

Overview of Optimization

Summary

This chapter defines optimization and its basic concepts. It provides examples of various engineering optimization problems.

1.1 Optimization

Engineers are commonly confronted with the tasks of designing and operating systems to meet or surpass specified goals while meeting numerous constraints imposed on the design and operation. Optimization is the organized search for such designs and operating modes. It determines the set of actions or elements that must be implemented to achieve optimized systems. In the simplest case, optimization seeks the maximum or minimum value of an objective function corresponding to variables defined in a feasible range or space. More generally, optimization is the search of the set of variables that produces the best values of one or more objective functions while complying with multiple constraints. A single-objective optimization model embodies several mathematical expressions including an objective function and constraints as follows:

$$Optimize\ f(X),\quad X=\left(x_1,x_2,\ldots,x_i,\ldots,x_N\right) \tag{1.1}$$

subject to

$$g_j(X)<b_j,\quad j=1,2,\ldots,m \tag{1.2}$$

$$x_i^{(L)}\le x_i\le x_i^{(U)},\quad i=1,2,\ldots,N \tag{1.3}$$

in which $f(X)$ = the objective function; X = a set of decision variables x_i that constitutes a possible solution to the optimization problem; x_i = ith decision variable; N = the number of decision variables that determines the dimension

Meta-Heuristic and Evolutionary Algorithms for Engineering Optimization,
First Edition. Omid Bozorg-Haddad, Mohammad Solgi, and Hugo A. Loáiciga.

of the optimization problem; $g_j(X)$ = jth constraint; b_j = constant of the jth constraint; m = the total number of constraints; $x_i^{(L)}$ = the lower bound of the ith decision variable; and $x_i^{(U)}$ = the upper bound of the ith decision variable.

1.1.1 Objective Function

The objective function constitutes the goal of an optimization problem. That goal could be maximized or minimized by choosing variables, or decision variables, that satisfy all problem constraints. The desirability of a set of variables as a possible solution to an optimization problem is measured by the value of objective function corresponding to a set of variables.

Some of the algorithms reviewed in this book are explained with optimization problems that involve maximizing the objective function. Others do so with optimization problems that minimize the objective function. It is useful to keep in mind that a maximization (or minimization) problem can be readily converted, if desired, to a minimization (or maximization) problem by multiplying its objective function by −1.

1.1.2 Decision Variables

The decision variables determine the value of the objective function. In each optimization problem we search for the decision variables that yield the best value of the objective function or optimum.

In some optimization problems the decision variables range between an upper bound and a lower bound. This type of decision variables forms a continuous decision space. For example, choosing adequate proportions of different substances to make a mixture of them involves variables that are part of a continuous decision space in which the proportions can take any value in the range [0,1]. On the other hand, there are optimization problems in which the decision variables are discrete. Discrete decision variables refer to variables that take specific values between an upper bound and a lower bound. Integer values are examples of discrete values. For instance, the number of groundwater wells in a groundwater withdrawal problem must be an integer number. Binary variables are of the discrete type also. The typical case is that when taken the value 1 implies choosing one type of action, while taking the value 0 implies that no action is taken. For example, a decision variable equal to 1 could mean building a water treatment plant at a site, while its value equal to 0 means that the plant would not be constructed at that site. Optimization problems involving continuous decision variables are called continuous problems, and those problems defined in terms of discrete decision variables are known as discrete problems. There are, furthermore, optimization problems that may involve discrete and continuous variables. One such example would be an optimization involving the decision of whether or not to build a facility at a certain location and, if so, what its capacity ought to be. The siting variable is of the binary type (0 or 1), whereas its capacity is a real, continuous variable. This type of optimization problem is said to be of mixed type.

1.1.3 Solutions of an Optimization Problem

Each objective function is expressed in terms of decision variables. When there is only one decision variable, the optimization problem is said to be one-dimensional, while optimization problems with two or more decision variables are called N-dimensional. An N-dimensional optimization problem has solutions that are expressed in terms of one or more sets of solutions in which each solution has N decision variables.

1.1.4 Decision Space

The set of decision variables that satisfy the constraints of an optimization problem is called the feasible decision space. In an N-dimensional problem, each possible solution is an N-vector variable with N elements. Each element of this vector is a decision variable. Optimization algorithms search for a point (i.e., a vector of decision variables) or points (i.e., more than one vector of decision variables) in the decision space that optimizes the objective function.

1.1.5 Constraints or Restrictions

Each optimization problem may have two types of constraints. Some constraints directly restrict the possible value of the decision variables, such as a decision variable x being a positive real number, $x > 0$, or analogous to Equation (1.3). Another form of constraint is written in terms of formulas, such as when two decision variables x_1 and x_2 are restricted to the space $x_1 + x_2 \leq b$ or analogous to Equation (1.2). The goal of an optimization problem is to find an optimal feasible solution that satisfies all the constraints and yields the best value of the objective function among all feasible solutions. Figure 1.1 depicts a constrained two-dimensional decision space with infeasible and feasible spaces.

The set of all feasible solutions constitute the feasible decision space, and the infeasible decision space is made up of all the infeasible decision variables. Evidently, the optimal solution must be in the feasible space.

1.1.6 State Variables

State variables are dependent variables whose values change as the decision variables change their values. State variables are important in engineering problems because they describe the system being modeled and the objective function and constraints are evaluated employing their values. As an example, consider an optimization problem whose objective is to maximize hydropower generation by operating a reservoir. The decision variable is the amount of daily water release passing through turbines. The state variable is the amount

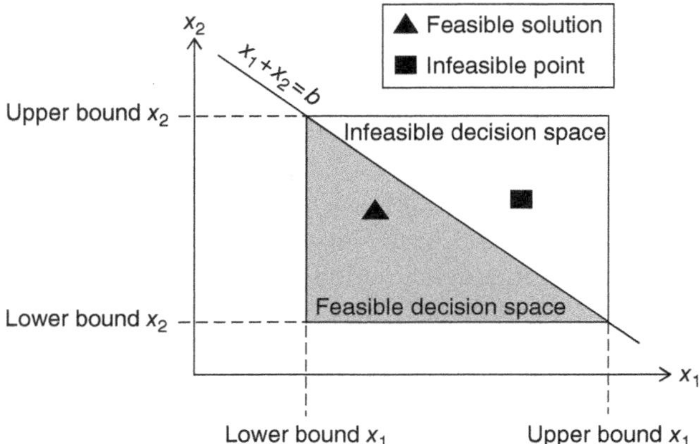

Figure 1.1 Decision space of a constrained two-dimensional optimization problem.

of water stored in the reservoir, which is affected by the water released through turbines according to an equation of water balance that also involves water inflow to the reservoir, evaporation from the reservoir, water diversions or imports to the reservoir, water released from the reservoir bypassing turbines, and other water fluxes that change the amount of reservoir storage.

1.1.7 Local and Global Optima

It has been established that a well-defined optimization problem has a well-defined decision space. Each point of the decision space defines a value of the objective function. A local optimum refers to a solution that has the best objective function in its neighborhood. In a one-dimensional optimization problem, a feasible decision variable X^* is a local optimum of a maximization problem if the following condition holds:

$$f(X^*) \geq f(X), \quad X^* - \varepsilon \leq X \leq X^* + \varepsilon \tag{1.4}$$

In a minimization problem the local-optimum condition becomes

$$f(X^*) \leq f(X), \quad X^* - \varepsilon \leq X \leq X^* + \varepsilon \tag{1.5}$$

where X^* = a local optimum and ε = limited length in the neighborhood about the local optimum X^*. A local optimum is limited to a neighborhood of the decision space, and it might not be the best solution over the entire decision space.

A global optimum is the best solution in the decision space. Some optimization problems may have more than one—in fact, an infinite number of global optima. These situations arise commonly in linear programming problems. In this case, all the global optima produce the same value of the objective function. There are not decision variables that produce a better objective function

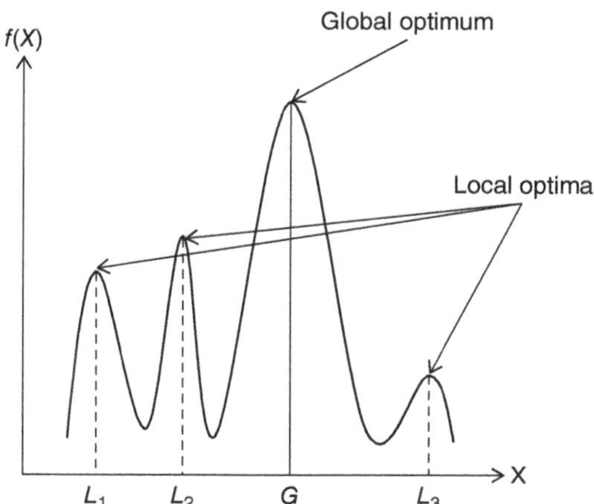

Figure 1.2 Schematic of global and local optimums in a one-dimensional maximizing optimization problem.

value than the global optimum. A one-dimensional optimization problem with decision variable X and objective function $f(X)$ the value X^* is a global optimum of a maximization problem if for any decision variable X the following is true:

$$f(X^*) \geq f(X) \tag{1.6}$$

In a minimization problem we would have

$$f(X^*) \leq f(X) \tag{1.7}$$

Figure 1.2 illustrates global and local optima for a one-dimensional maximization problem.

L_1, L_2, and L_3 in Figure 1.2 are local optima, and G denotes the global optimum with the largest value of the objective function. The decision space may be single modal or multimodal. In a single-modal surface, there is only one extreme point, while there are several extremes on the surface of a multimodal problem. In a single-modal problem, there is a single local optimum that is also the global optimum. On the other hand, a multimodal problem may include several local and global optima. However, the decision variables that produce a global optimum must all produce the same value of the global optimum, by definition. Figure 1.3 illustrates the surface of one-dimensional optimization problems with single-modal and multimodal decision spaces in which there is one single optimum.

1.1.8 Near-Optimal Solutions

A near optimum has a very close but inferior value to the global optimum. In some engineering problems, achieving the absolute global optimum is extremely difficult or sometimes impossible because of the innate complexity of the problem or the method employed to solve the problem. Or achieving the

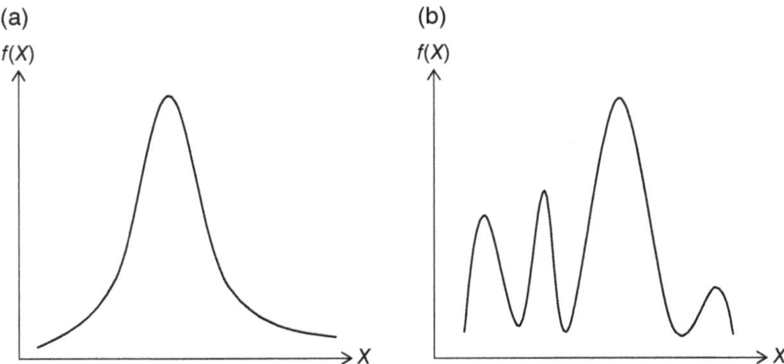

Figure 1.3 Different types of decision spaces: (a) maximization problem with single-modal surface and one global optimum; (b) maximization problem with multimodal surface that has one global optimum.

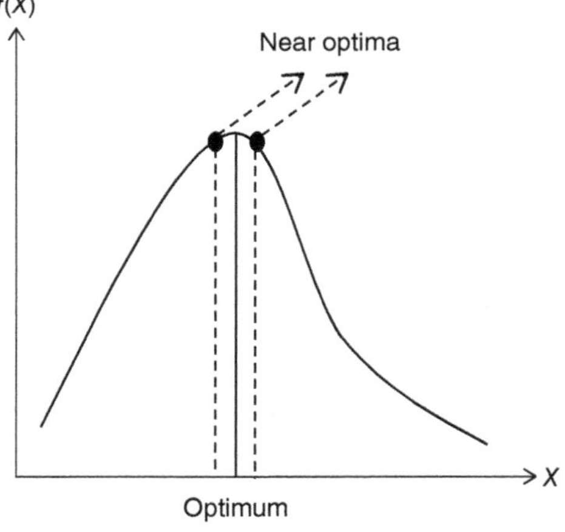

Figure 1.4 Demonstration of near optima in a one-dimensional maximizing optimization problem.

global optimum may be computationally prohibitive. In this situation, a near optimum is calculated and reported as an approximation to the global optimum. Near optima are satisfactory in solving many real-world problems. The proximity of a near optimum to the global optimum depends on the optimization problem being solved and the judgment of the analyst. Figure 1.4 depicts the concept of a near optimum in a maximization problem.

1.1.9 Simulation

Each decision variable of an optimization problem defines an objective function value. The process of evaluating the state variables, which are necessary for estimation of the objective function, and constraints with any decision variable is known as simulation. A simulation model receives the decision variables as inputs and simulates the system's state variables. Sometimes the simulation model consists of one or more simple mathematical functions and equations. However, most real-world and engineering problems require simulation models with complex procedures that most solve systems of equations and various formulas that approximate physical processes. Simulation is, therefore, the computational imitation of the operation of a real-world process or system over time.

1.2 Examples of the Formulation of Various Engineering Optimization Problems

This section presents examples of the formulation of different types of engineering optimization problems including mechanical design, structural design, electrical engineering optimization, water resources optimization, and calibration of hydrological models.

1.2.1 Mechanical Design

Designing a compound gear train is exemplary of optimal designing. A compound gear train is designed to achieve a particular gear ratio between the driver and driven shafts. The purpose of the gear train design is finding the number of teeth in each gear so that the error between the obtained and required gear ratios is minimized. In practice, the term gear ratio is used interchangeably with velocity ratio. It is defined as the ratio of the angular velocity of the output shaft to that of the input shaft. For a pair of matching gears, the velocity or gear ratio α is calculated as follows:

$$\alpha = \left|\frac{\omega_{out}}{\omega_{in}}\right| = \frac{\theta_{in}}{\theta_{out}} \tag{1.8}$$

in which α = gear ratio; ω_{in} = angular velocity of the input shaft; ω_{out} = angular velocity of the output shaft; θ_{in} = the number of teeth on the input gear; and θ_{out} = the number of teeth on the output gear. The ratio is, thus, inversely proportional to the number of teeth on the input and output gears.

Figure 1.5 shows a compound gear train that is made of four gears. It is desired to produce a gear ratio as close as possible to a required value μ. The objective of the design is to find the number of teeth in each gear so that the error between the obtained and required gear ratios is minimized. Normally, additional considerations such as the number of gear pairs to use and the geometric arrangement of the shafts must be considered in addition to wear. To simplify the problem only the particular configuration shown in Figure 1.5 is considered here.

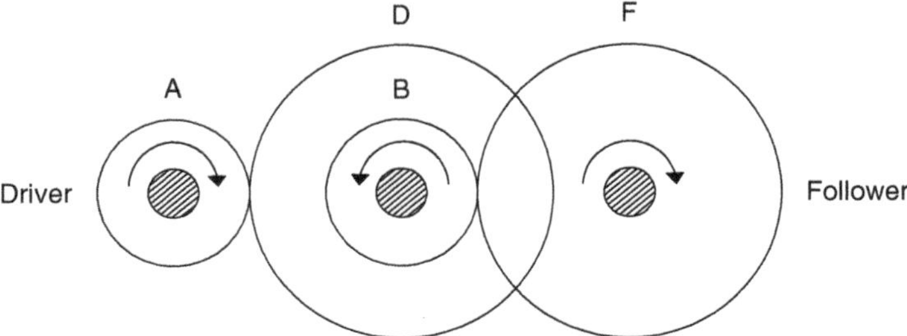

Figure 1.5 Compound gear train made of four gears.

For the system shown in Figure 1.5, the gear ratio is evaluated as follows:

$$\alpha = \frac{\tau_d}{\tau_a} \times \frac{\tau_b}{\tau_f} \tag{1.9}$$

in which τ_d, τ_a, τ_b, and τ_f = the number of teeth on gears D, A, B, and F, respectively.

The number of teeth on each gear constitutes the decision variables:

$$X = (x_1, x_2, x_3, x_4) = (\tau_d, \tau_a, \tau_b, \tau_f) \tag{1.10}$$

Minimizing the square of the difference between the desired gear ratio (μ) and the actual design gear ratio (α) the optimization problem leads to the following optimization problem:

$$Minimize\ f(X) = (\mu - \alpha)^2 \tag{1.11}$$

in which

$$\alpha = \frac{x_1}{x_2} \times \frac{x_3}{x_4} \tag{1.12}$$

subject to

$$x^{(L)} \le x_i \le x^{(U)}, \quad i = 1, 2, 3, 4 \tag{1.13}$$

where μ = required gear ratio; α = actual gear ratio; $x^{(L)}$ and $x^{(U)}$ = minimum and maximum number of teeth on each gear, respectively. The minimization of the objective function (Equation (1.11)) is with respect to x_1, x_2, x_3, and x_4. The objective function is nonlinear, and the constraints (Equation (1.13)) are simple bounds on the decision variables. Since the number of teeth is an integer number, this problem has a discrete domain, and the decision variables must take integers values.

1.2.2 Structural Design

Structural optimization problems are created and solved to determine the configurations of structures that satisfy specifications and produce an optimum for a chosen objective function. The main purpose of structural optimization is to minimize the weight of a structure or the vertical deflection of a loaded member. Here, a two-bar truss design model is considered for illustration purposes.

The truss shown in Figure 1.6 is designed to carry a certain load without elastic failure. In addition, the truss is subject to limitations in geometry, area, and stress.

The stresses on nodes A and B are calculated as follows:

$$\sigma_{AC} = \frac{Force \times \left(\frac{L-L'}{L}\right) \times \sqrt{H^2 + L'^2}}{a_1 \times H} \tag{1.14}$$

$$\sigma_{BC} = \frac{Force \times \left(\frac{L'}{L}\right) \times \sqrt{H^2 + (L-L')^2}}{a_2 \times H} \tag{1.15}$$

in which σ_{AC} and σ_{BC} = the stress on node A and B, respectively (N/m^2); *Force* = force on node C (N); H = perpendicular distance from AB to point C (m); L = length of AB (m); L' = length of AC (m); a_1 = cross-sectional area of AC; and a_2 = cross-sectional area of BC (m^2).

In this case, a_1, a_2, and H are the decision variables of the optimization model:

$$X = (x_1, x_2, x_3) = (a_1, a_2, H) \tag{1.16}$$

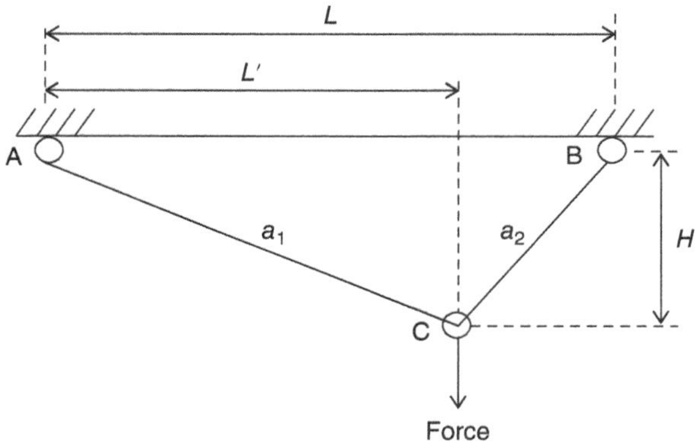

Figure 1.6 Schematic of a two-bar truss.

The optimization problem is expressed as follows when the weight of the structure is minimized:

$$Minimize\ f(X) = \rho \times \left(\left(a_1 \times \sqrt{H^2 + L^2} \right) + \left(a_2 \times \sqrt{H^2 + (L - L')^2} \right) \right) \quad (1.17)$$

subject to

$$\sigma_{AC}, \sigma_{BC} \leq \sigma_{max} \quad (1.18)$$

$$x_i^{(L)} \leq x_i \leq x_i^{(U)}, \quad i = 1,2,3 \quad (1.19)$$

in which ρ = the volumetric density of the truss; σ_{max} = the maximum allowable stress; $x_i^{(L)}$ and $x_i^{(U)}$ = the minimum and maximum values of the decision variables, respectively. The minimization of Equation (1.17) is with respect to a_1, a_2, and H, which are real-valued variables. The objective function is nonlinear, and so are the constraints.

1.2.3 Electrical Engineering Optimization

Directional overcurrent relays (DOCRs), which protect transmission systems, constitute a classical electrical engineering design problem. DOCRs are part of electrical power systems that isolate faulty lines in the event of failures in the system. DOCRs are logical elements that issue a trip signal to the circuit breaker if a failure occurs within the relay jurisdiction and are placed at both ends of each transmission line. Their coordination is an important aspect of system protection design. The relay coordination problem is to determine the sequence of relay operations for each possible failure location so that the failure section is isolated with sufficient coordination margins and without excessive time delays. The selection of the sequence of relay operations is a function of the power network topology, relay characteristics, and protection philosophy. The DOCR protection scheme consists of two types of settings, namely, current, referred to as plug setting (PS), and time dial setting (TDS), which must be calculated. With the optimization of these settings, an efficient coordination of relays can be achieved, and the faulty transmission line may be isolated, thereby maintaining a continuity of power supply to functional sections of power systems.

The operating time (T) of a DOCR is a nonlinear function of the relay settings including time dial settings (TDS), plug settings (PS), and the fault current (I) seen by the relay. The relay operating time equation for a DOCR is estimated as follows:

$$T = \frac{K_1 \times \xi}{\left(\frac{\gamma}{\psi \times C_{pri-rating}} \right)^{K_2} - K_3} \quad (1.20)$$

in which T = operating time; K_1, K_2, and K_3 = constants that depend upon the specific device being simulated; ξ = time dial settings; ψ = plug settings; γ = faulty current passing through the relay, which is a known value, as it is a system-dependent parameter and continuously measured by monitoring instruments; and $C_{pri-rating}$ = a parameter whose value depends on the number of turns in the current transformer (CT). CT is used to reduce the level of the current so that the relay can withstand it. One “current transformer” is used with each relay, and, thus, $C_{pri-rating}$ is known in the problem.

The TDS and PS of the relays are the decision variables of the optimization model:

$$X=\left(x_1,x_2,\ldots,x_N,x_{N+1},x_{N+2},\ldots,x_{2N}\right)=\left(\xi_1,\xi_2,\ldots,\xi_N,\psi_1,\psi_2,\ldots,\psi_N\right) \tag{1.21}$$

where N = number of relays of the system.

The optimization problem is formulated as follows:

$$Minimize\ f(X)=\sum_{i=1}^{N}\sum_{j=1}^{M}T_{i,j}^{(primary)} \tag{1.22}$$

subject to

$$T^{(backup)}-T^{(primary)}\geq C_T,\quad for\ all\ relay\ pairs \tag{1.23}$$

$$x_i^{(L)}\leq x_i\leq x_i^{(U)},\quad i=1,2,\ldots,2N \tag{1.24}$$

in which M = number of failures; $T_{i,j}^{(primary)}$ = operating time of the primary relay i for a failure j; $T^{(backup)}$ = operating time of the backup relay; C_T = coordinating time interval; and $x_i^{(L)}$ and $x_i^{(U)}$ = bounds on relay settings. The objective function (Equation (1.22)) is nonlinear, and so are the constraints.

1.2.4 Water Resources Optimization

Flowing water generates energy that can be managed and turned into electricity. This is known as hydroelectric power or hydropower. Dams of reservoirs are the most common type of hydroelectric power plant. Some hydropower dams have several functions such as supplying urban and agriculture water and flood control. This example focuses on hydropower generation exclusively. Figure 1.7 shows a simple schematic of a hydropower dam with its associated water fluxes.

The reservoir storage (S) in each operational time period is calculated as follows:

$$S_{t+1}=I_t+S_t-R_t-Sp_t,\quad t=1,2,\ldots,N \tag{1.25}$$

$$Sp_t=\begin{cases}0 & if\ I_t+S_t-R_t\leq S_{max}\\ I_t+S_t-R_t-S_{max} & if\ I_t+S_t-R_t>S_{max}\end{cases} \tag{1.26}$$

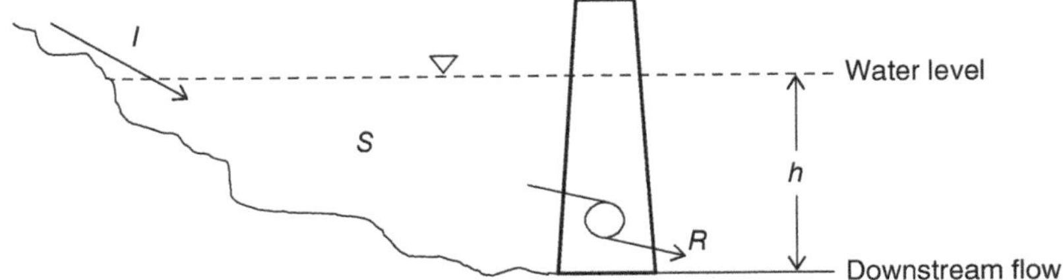

Figure 1.7 Schematic of a hydropower dam.

in which S_{t+1} = the reservoir storage at the end of time period t; I_t = the volume of input water during time period t; S_t = the reservoir storage at the start of time period t; R_t = the volume of release of the reservoir; S_{max} = the reservoir capacity; Sp_t = the volume of overflow that occurs whenever the reservoir storage exceeds the reservoir capacity; and N = the total number of time periods. There is spill or overflow whenever the reservoir storage exceeds the capacity of the reservoir. The storage volume cannot be less than a value like S_{min} given that the floodgate is usually placed higher than the bottom of the dam. This part of the reservoir is usually filed with sediments.

The generated power is a function of water flow and the elevation difference between the hydraulic head at the intake and outlet of the turbine. The generated power in period t is determined as follows:

$$P_t = \eta \times \rho \times g \times h_t \times q_t \tag{1.27}$$

$$q_t = \frac{R_t}{\Delta t \times n_t} \tag{1.28}$$

in which P_t = generated power (W) in period t; η = efficiency of powerhouse; ρ = density (kg/m^3) (~1000 kg/m^3 for water); g = acceleration of gravity (9.81 m/s^2); h_t = falling height or effective hydraulic head (m) in time period t; q_t = water flow in period t (m^3/s); R_t = total volume of release of reservoir in time period t (m^3); n_t = ratio of time when the powerhouse is active; and Δt = length of time period (s). The water level in the reservoir is usually estimated based on the area or volume of water stored in the reservoir by predefined equations.

The volume of water that enters the reservoir in each time period (I_t) is known. The goal of the problem is determining the volume of release in each time period (R_t) so that the total generated power is close to the power plant capacity (PPC) as much as possible. The volume of water release from the reservoir in each time period is the decision variable of the optimization model:

$$X = (x_1, x_2, \ldots, x_N) = (R_1, R_2, \ldots, R_N) \tag{1.29}$$

where N = total number of decision variables that is equal to the total number of time periods.

The optimization problem minimizes the normalized sum of squared deviations between generated power and power plant capacity and is written as follows:

$$Minimize\ f(X)=\sum_{i=1}^{N}\left(1-\frac{P_i}{PPC}\right)^2 \tag{1.30}$$

$$P_i=\eta\times\rho\times g\times h_i\times\frac{x_i}{\Delta t\times n_i},\quad i=1,2,\ldots,N \tag{1.31}$$

subject to

$$S_{min}\le S_i\le S_{max} \tag{1.32}$$

$$0\le P_i\le PPC \tag{1.33}$$

$$x_i^{(L)}\le x_i\le x_i^{(U)},\quad i=1,2,\ldots,N \tag{1.34}$$

in which PPC = the power plant capacity; S_{min} = the minimum storage; and $x_i^{(L)}$ and $x_i^{(U)}$ = the minimum and maximum volume of release in each period i, respectively. The minimum release is governed by different factors such as environmental flows to sustain aquatic life. One of the factors that restrict the maximum release from the reservoir is the capacity of the floodgate.

1.2.5 Calibration of Hydrologic Models

The parameter calibration of hydrologic model is commonly posed as the minimization of a norm of errors between observed and predicted hydrologic values. The routing of floods in river channels is a classic example involving the calibration of hydrologic models. Flood is a natural phenomenon that can cause considerable damage in urban, industrial, and agricultural regions. To prevent those damages it is necessary to implement a hydrologic model to estimate the flood hydrograph at the downstream river reach given the upstream hydrograph. The Muskingum model is a hydrologic model based on the continuity and parameterized storage equations as follows:

Continuity:

$$\frac{dS_t}{dt}=I_t-O_t \tag{1.35}$$

Parameterized storage:

$$S_t=\beta_1\left[Y\times I_t+(1-Y)O_t\right]^{\beta_2} \tag{1.36}$$

in which S_t, I_t, and O_t = storage, inflow, and outflow in a river reach at time t, respectively; β_1 = storage time constant for a river reach that has a value reasonably close to the travel time of a flood through the river reach; β_2 = exponent for the effect of nonlinearity between accumulated storage and weighted flow;

and Y = weighting factor between 0 and 0.5 for reservoir storage and between 0 and 0.3 for stream channels.

Equation (1.36) is solved for the reach outflow as follows:

$$O_t = \left(\frac{1}{1-Y}\right)\left(\frac{S_t}{\beta_1}\right)^{\frac{1}{\beta_2}} - \left(\frac{Y}{1-Y}\right)I_t \tag{1.37}$$

By combining Equations (1.35) and (1.37), the state equation becomes

$$\frac{\Delta S_t}{\Delta t} = -\left(\frac{1}{1-Y}\right)\left(\frac{S_t}{\beta_1}\right)^{\frac{1}{\beta_2}} + \left(\frac{1}{1-Y}\right)I_t \tag{1.38}$$

$$S_{t+1} = S_t + \Delta S_t \tag{1.39}$$

The routing of a flood hydrograph consists of the following steps:

Step 1: Assume values for the parameters β_1, β_2, and Y.
Step 2: Calculate the storage (S_t) with Equation (1.36), with the initial outflow equal to the initial inflow.
Step 3: Calculate the time rate of change of storage volume with Equation (1.38).
Step 4: Estimate the next storage with Equation (1.39).
Step 5: Calculate the next outflow with Equation (1.37).
Step 6: Repeat steps 2–5 for total time steps in the flood routing until reaching a stopping criterion.

The goal of this problem is estimating β_1, β_2, and Y so that the sum of the squared differences between observed and predicted outflows is minimized. The parameters β_1, β_2, and Y of the Muskingum model are the decision variables:

$$X = (x_1, x_2, x_3) = (\beta_1, \beta_2, Y) \tag{1.40}$$

The optimization problem is formulated as follows:

$$\text{Minimize } SSQ = \sum_{t=1}^{M}\left(O_t - \hat{O}_t\right)^2 \tag{1.41}$$

subject to

$$x_i^{(L)} \le x_i \le x_i^{(U)}, \quad i = 1,2,3 \tag{1.42}$$

where M = total number of time steps in the flood routing; O_t and $\hat{O}_t$ = observed and routed outflow, respectively, at time t; and $x_i^{(L)}$ and $x_i^{(U)}$ = minimum and maximum values of parameters of Muskingum model, respectively.

1.3 Conclusion

This chapter introduced foundational concepts of optimization such as the objective function, decision variables, decision space, and constraints. In addition, several examples of formulating engineering optimization problems were presented to illustrate a variety of optimization models.

2

Introduction to Meta-Heuristic and Evolutionary Algorithms

Summary

This chapter presents a brief review of methods for searching the decision space of optimization problems, describes the components of meta-heuristic and evolutionary algorithms, and illustrates their relation to engineering optimization problems. Other topics covered in this chapter are the coding of meta-heuristic and evolutionary algorithms, dealing with constraints, the generation of initial or tentative solutions, the iterative selection of solutions, and the performance evaluation of meta-heuristic and evolutionary algorithms. A general algorithm that encompasses the steps of all meta-heuristic and evolutionary algorithms is presented.

2.1 Searching the Decision Space for Optimal Solutions

The set of all possible solutions for an optimization problem constitutes the decision space. The goal of solving an optimization problem is finding a solution in the decision space whose value of the objective function is the best among all possible solutions. One of the procedures applicable for finding the optimum in a decision space is sampling or trial-and-error search. The methods that apply trial-and-error search include (1) sampling grid, (2) random sampling, and (3) targeted sampling.

The goal of a sampling grid is evaluating all possible solutions and choosing the best one. If the problem is discrete, a sampling network evaluates all possible solutions and constraints. The solution that satisfies all the constraints and has the best objective function value among all feasible solutions is chosen as the optimum. When the decision space of a discrete problem is large, the computational burden involved in evaluating the objective

Meta-Heuristic and Evolutionary Algorithms for Engineering Optimization,
First Edition. Omid Bozorg-Haddad, Mohammad Solgi, and Hugo A. Loáiciga.

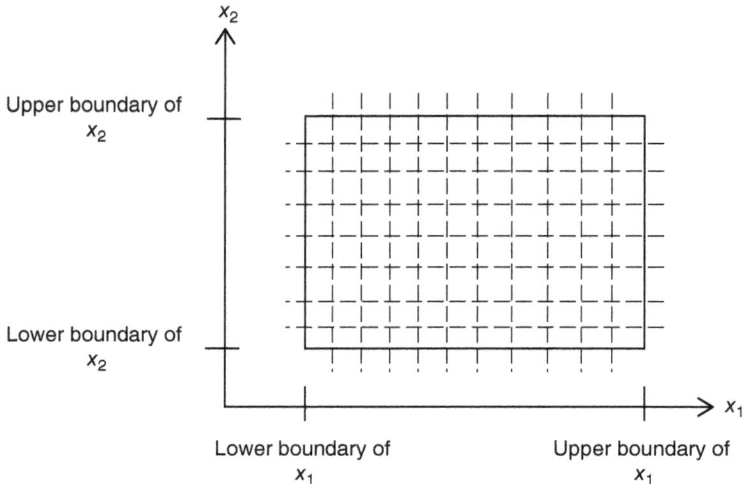

Figure 2.1 Sampling grid on a two-dimensional decision space.

function and constraints could be prohibitive. Therefore, the sampling grid method is practical for relatively small problems only. When an optimization problem is continuous testing, all solutions are not possible because there are an infinite number of them. In this situation the continuous problem is transformed to a discrete problem by overlaying a grid on the decision space as shown in Figure 2.1. The intersections of the grid are points that are evaluated. In fact, after discretization of the decision space, the procedure followed is as the same as that employed for discrete problems. It is clear that in this method reducing the size of the grid interval improves the accuracy of the search while increasing the computational burden. It is generally impossible to find solutions that are very near the global optimum of a complex optimization problem because to achieve that it would be necessary to choose a very small grid for which the computational burden would in all probability be prohibitive.

Another method is random sampling in which sampling is performed randomly through the decision space. Possible solutions are chosen randomly and their objective functions are evaluated. The best solution among the chosen possible solutions is designated as the optimum.

Suppose that there are S possible solutions, among which $r = 1$ is the optimal one, and K possible solutions are chosen randomly among the S possible ones to be evaluated. First, let us consider that the random selection is done without replacement, and let Z denote the number of optimal solutions found in the randomly chosen sample of K possible solutions (Z can only take the value 0 or 1 in this instance). The probability that one of the K chosen possible solutions is the optimal one is found from the hypergeometric distribution and is equal to $P(Z=1)=K/S$. Therefore, if there are $S=10^6$ possible solutions and $K=10^5$ possible solutions are randomly chosen, the probability of selecting the optimal solution among those in the randomly chosen sample is only 0.10 (10.0%) in spite of the computational effort of evaluating 10^5 possible solutions. Also, random selection can be done with replacement. In this method, the probability that one of the tested solutions is the optimal solution of the optimization problem equals $P(Z=1)=1-((S-1)/S)^K=1-(999999/10^6)^{10^5}=0.095$ or about 9.5%.

One of the key shortcomings of the previous sampling grid and random sampling methods is that they require that all the decision space be searched precisely. This exerts a high and wasteful computational effort. In these two methods, the evaluation of any new possible solution is done independently of previously tested solutions. In others words, there is no learning about the history of previous computations to guide the search for the optimal solution more efficiently as the search algorithm progresses through the computations.

The sampling grid and random sampling are not efficient or practical methods to solve real-world engineering problems, and they are cited here as an introduction to a third method called targeted sampling. Unlike sampling grid and random sampling, targeted sampling searches the decision space, taking into account the knowledge gained from previously tested possible solutions, and selects the next sample solutions based on results from previously tested solutions. Thus, targeted sampling focuses gradually in areas of the decision space where the optimum may be found with a high probability.

Targeted sampling is the basis of all meta-heuristic and evolutionary algorithms that rely on a systematic search to find an optimum. In contrast to other sampling methods, meta-heuristic and evolutionary algorithms of the targeted sampling type are capable to solve all well-posed real-world and complex problems that other types of optimization methods such as linear and nonlinear programming, dynamic programming, and stochastic dynamic programming cannot solve. For this reason meta-heuristic and evolutionary algorithms have become a preferred solution approach for most complex engineering optimization problems.

Meta-heuristic and evolutionary algorithms are typically applied to calculate near-optimal solutions of problems that cannot be solved easily or at all using other techniques, which constitute the great majority of problems. Meta-heuristic and evolutionary algorithms may prove to be computationally intensive in finding an exact solution, but sometimes a near-optimal solution is sufficient. In these situations, evolutionary techniques are effective. Due to their random search nature, evolutionary algorithms are never guaranteed to find an optimal solution to any problem, but they will often find a near-optimal solution if one exists.

2.2 Definition of Terms of Meta-Heuristic and Evolutionary Algorithms

Meta-heuristic and evolutionary algorithms are problem-independent techniques that can be applied to a wide range of problems. An “algorithm” refers to a sequence of operations that are performed to solve a problem. Algorithms are made of iterative operations or steps that are terminated when a stated convergence criterion is reached. Each step may be refined into more refined detail in terms of simple operations. Figure 2.2 shows a general schematic of an algorithm.

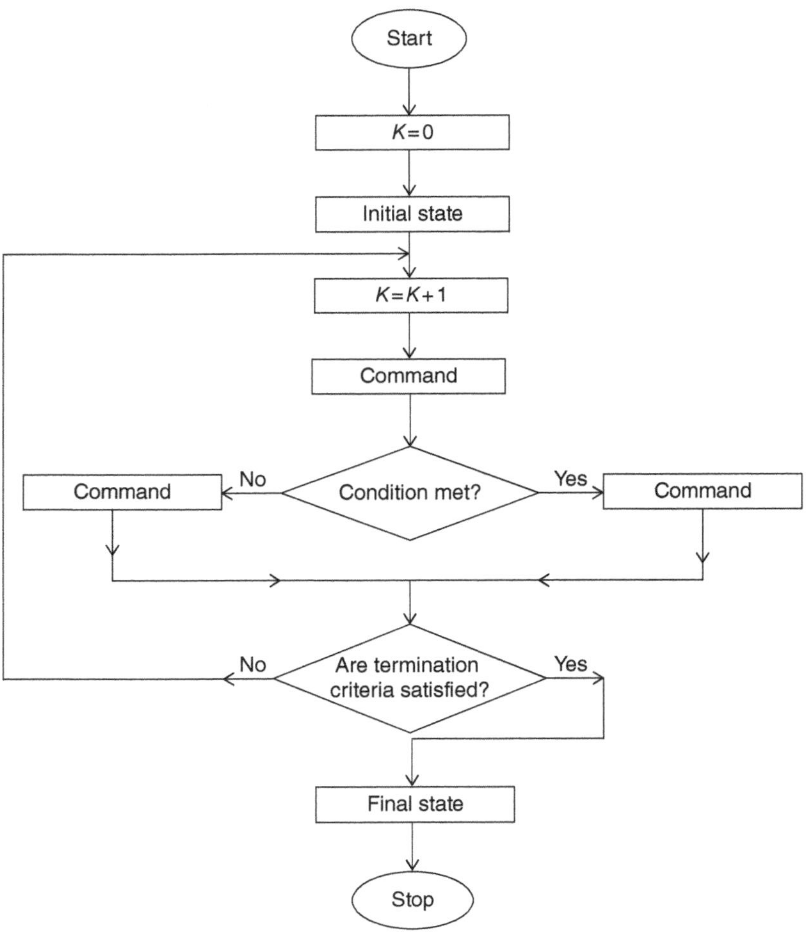

Figure 2.2 General schematic of a simple algorithm; K denotes the counter of iterations.

Meta-heuristic and evolutionary algorithms start from an initial state and initial data. The purpose of these algorithms is finding appropriate values for the decision variables of an optimization problem so that the objective function is optimized. Although there are differences between meta-heuristic and evolutionary algorithms, they all require initial data and feature an initial state, iterations, final state, decision variables, state variables, simulation model, constraints, objective function, and fitness function.

2.2.1 Initial State

Each meta-heuristic and evolutionary algorithm starts from an initial state of variables. This initial state can be predefined, randomly generated, or deterministically calculated from formulas.

2.2.2 Iterations

Algorithms perform operations iteratively in the search for a solution. Evolutionary or meta-heuristic algorithms start their iterations with one or several initial solutions of the optimization problem. Next, sequential operations are performed to generate new solution(s). An iteration ends when a new possible solution is generated. The new generated solution(s) is (are) considered as initial solution(s) for the next iteration of the algorithm.

2.2.3 Final State

After satisfying the chosen termination criteria, the algorithm stops and reports the best or final generated solution(s) of an optimization problem. Termination criteria are defined in several different forms: (1) the number of iterations, (2) the improvement threshold of the value of solution between consecutive iterations, and (3) the run time of the optimization algorithm. The first criterion refers to a predefined number of iterations that the algorithm is allowed to execute. The second criterion sets a threshold for improving the solution between consecutive steps. The third criterion stops the algorithm after a defined run time and the best solution available at that time is reported.

2.2.4 Initial Data (Information)

Initial information is classified into two categories including (1) data about the optimization problem, which are required for simulation, and (2) parameters of the algorithm, which are required for its execution and may have to be calibrated. Table 2.1 lists the input data for several sample problems introduced in Chapter 1.

The second type of initial information is needed to calibrate the solution algorithm to solve an optimization problem. Almost all meta-heuristic and evolutionary algorithms have parameters that must be adjusted. The parameters of the algorithm must be properly chosen to achieve its successful application. The stopping criterion, for instance, is an algorithmic parameter

Table 2.1 The input data of the example problems presented in Chapter 1.

Problem	Input data
Mechanical design (compound gear train)	(1) The required gear ratio (2) Limits on decision variables
Structural design (two-bar truss)	(1) The distance between supports (2) The horizontal distance between loaded force and supports (3) The maximum allowable stress (4) Volume density (5) Load force (6) Limits on decision variables
Electrical engineering optimization (DOCR)	(1) Characteristics of the simulated system (2) Parameter of the current transformer (3) The number of faults (4) The number of relays (5) Parameter of fault current (6) Limits on decision variables
Water resources optimization (hydropower plant)	(1) Reservoir inflow (2) The reservoir capacity (3) The dead volume of the reservoir (4) The efficiency of the powerhouse (5) The density of water (6) Acceleration of gravity (7) The number of time periods (8) The length of the time periods (9) Power plant capacity (10) Limits on decision variables
Calibration of hydrologic models (Muskingum model)	(1) Number of time steps (2) Length of time steps (3) Reach inflow (4) Limits on decision variables

that is user specified. If the stopping criterion is not correctly chosen, the algorithm may not converge to the global solution. On the other hand, the algorithm could run for an unnecessarily long time.

2.2.5 Decision Variables

Decision variables are those whose values are calculated by execution of the algorithm, and their values are reported as solution of an optimization problem upon reaching the stopping criterion. Table 2.2 lists the decision variables for the problems introduced in Chapter 1.

Meta-heuristic and evolutionary algorithms first initialize the decision variables and recalculate their values through the execution of the algorithm.

2.2.6 State Variables

The state variables are related to the decision variables. In fact, the values of the state variables change as the decision variables change. Table 2.3 lists the state variables for the example problems introduced in Chapter 1.

2.2.7 Objective Function

The objective function determines the optimality of solutions. An objective function value is assigned to each solution of an optimization problem. Table 2.4 lists the objective functions of the example problems introduced in Chapter 1.

Table 2.2 The decision variables of the example problems presented in Chapter 1.

Problem	Decision variable
Mechanical design (compound gear train)	The number of tooth of the gears
Structural design (two-bar truss)	The properties of the truss
Electrical engineering optimization (DOCR)	The relay settings
Water resources optimization (hydropower plant)	The volume of water releases from the reservoir
Calibration of hydrologic models (Muskingum model)	The parameters of the Muskingum model

Table 2.3 The state variables of the example problems presented in Chapter 1.

Problem	State variable
Mechanical design (compound gear train)	The gear ratio
Structural design (two-bar truss)	Stress on nodes of the truss
Electrical engineering optimization (DOCR)	The operating time
Water resources optimization (hydropower plant)	Reservoir storage and generated power
Calibration of hydrologic models (Muskingum model)	Routed outflow and reach storage

Table 2.4 The objective function of the example problems presented in Chapter 1.

Problem	Objective function
Mechanical design (compound gear train)	Minimizing the differences between the required and calculated gear ratios
Structural design (two-bar truss)	Minimizing the weight of the structure
Electrical engineering optimization (DOCR)	Minimizing the summation of the operating times
Water resources optimization (hydropower plant)	Minimizing the differences between the power plant capacity and the generated power
Calibration of hydrologic models (Muskingum model)	Minimizing the differences between the observed and the routed outflows

2.2.8 Simulation Model

A simulation model is a single function or a set of mathematical operations that evaluate the values of the state variables in response to the values of the decision variables. The simulation model is a mathematical representation of a real problem or system that forms part of an optimization problem. The mathematical representation is in terms of numerical and logical operations programmed in the solution algorithm implemented for an optimization problem.

2.2.9 Constraints

Constraints delimit the feasible space of solutions of an optimization problem and are considered in meta-heuristic and evolutionary algorithms. In fact, these influence the desirability of each possible solution. After objective function and state variables related to each solution are evaluated, the constraints are calculated and define conditions that must be satisfied for feasibility of any possible solution. If the constraints are satisfied, the solution is accepted and it is called a feasible solution; otherwise the solution is removed or modified. Table 2.5 lists the constraints of the example problems introduced in Chapter 1.

2.2.10 Fitness Function

The value of the objective function is not always the chosen measure of desirability of a solution. For example, the algorithm may employ a transformed form of the objective function by the addition of penalties that avoid the violation of constraints, in which case the transformed function is called the fitness function. The fitness function is then employed to evaluate the desirability of possible solutions.

Table 2.5 The constraints of the example problems presented in Chapter 1.

Problem	Constraint
Mechanical design (compound gear train)	(1) Limits on the number of tooth of the gears
Structural design (two-bar truss)	(1) Limitation of the stress on nodes (2) Limits on the cross sectional area of trusses (3) Limits on the perpendicular distance between the load force and the supports
Electrical engineering optimization (DOCR)	(1) Limitation of the operating time of the primary and backup relays (2) Limits on the relay settings
Water resources optimization (hydropower plant)	(1) Limitation of the reservoir storage (2) Limitation of the generated power (3) Limits on the water release from reservoir
Calibration of hydrologic models (Muskingum model)	(1) Limits on the parameters of the Muskingum model

2.3 Principles of Meta-Heuristic and Evolutionary Algorithms

Figure 2.3 depicts the relation between the simulation model and the optimization algorithm in an optimization problem. The decision variables are inputs to the simulation model. Then, the state variables, which are outputs of the simulation model, are evaluated. Thereafter, the objective function is evaluated. In the next step, the problem constraints are evaluated, and lastly the fitness value of the current decision variables is calculated. At this time, the optimization algorithm generates a new possible solution of decision variables to continue the iterations if a termination criterion is not reached. Notice that if the optimization generates a set of solutions rather than a single solution in each iteration, the previous steps are performed for all solutions in parallel with each other. The meta-heuristic and evolutionary algorithms are independent of the simulation model and they only employ the value of the current state variables. In other words, these algorithms execute their operations independently of the equations and calculations executed by the simulation model. The main difference between the various meta-heuristic and evolutionary algorithms is how they generate new solution(s) in their iterative procedure, wherein these apply elements of artificial intelligence by learning from previous experience (old possible solutions) and employ accumulated information to generate new possible solutions. In other words, optimization algorithms generate a set of solutions whose fitness values are evaluated. Based on these fitness values, the optimization algorithm generates a new and improved set of solutions.

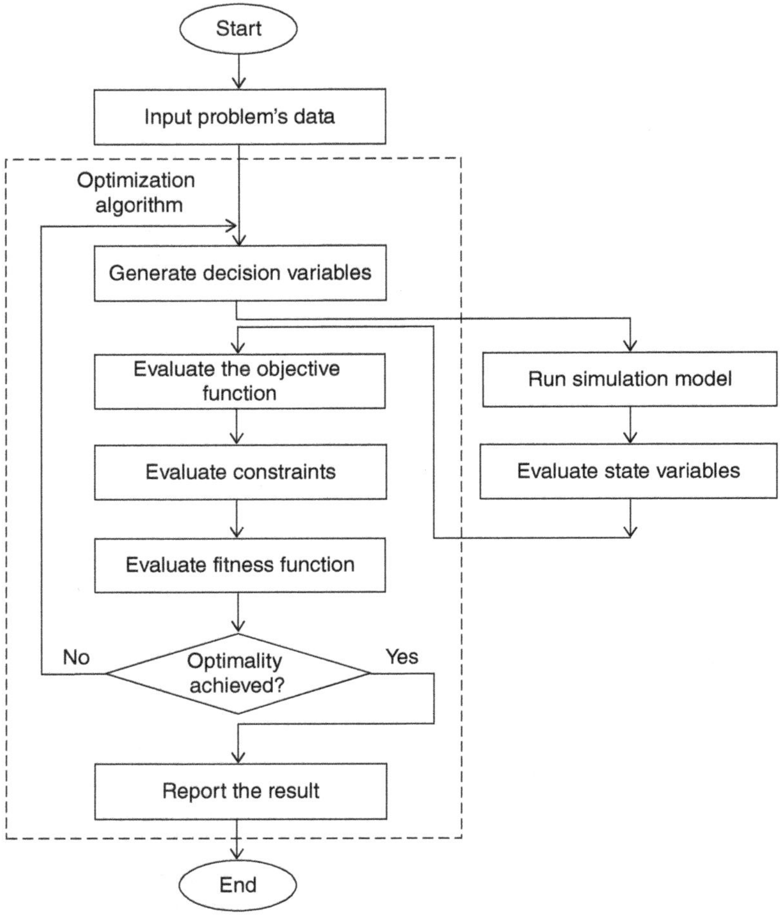

Figure 2.3 Diagram depicting the relation between a simulation model and an optimization algorithm.

In summary, meta-heuristic and evolutionary algorithms first generate a set of initial solutions. The simulation model then calculates the decision variables (these are the current possible solutions) with which to evaluate the objective function. The fitness values corresponding to the current decision variables are evaluated based on the calculated objective function. At this juncture the optimization algorithm applies a number of operations akin to phenomena observed in nature or that might be based on other principles to generate new solutions while it takes advantage of the good features of the previous solution(s). Optimization algorithms attempt to improve solutions in each

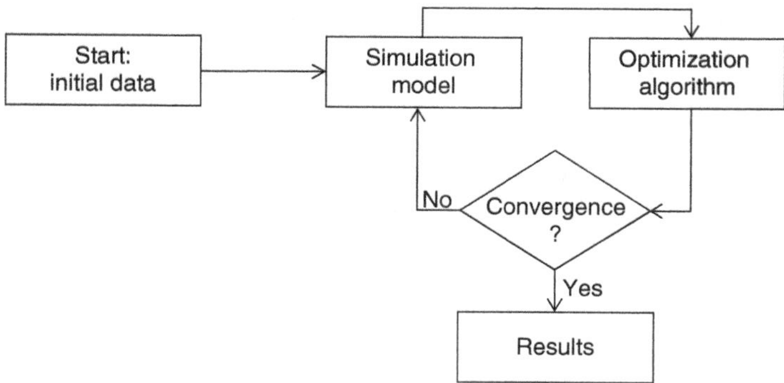

Figure 2.4 The main components of the optimization by meta-heuristic and evolutionary algorithms.

iteration, seeking to converge toward the optimal solution. After a number of iterations, the search reaches an optimal region of the feasible decision space. The best solution calculated by the algorithm at the time of termination constitutes the optimal solutions of a particular run. Figure 2.4 portrays the process of optimization by meta-heuristic and evolutionary algorithms.

2.4 Classification of Meta-Heuristic and Evolutionary Algorithms

This section presents several classifications of meta-heuristic and evolutionary algorithms.

2.4.1 Nature-Inspired and Non-Nature-Inspired Algorithms

Some algorithms are inspired by natural process, such as the genetic algorithm (GA), the ant colony optimization (ACO), the honey-bee mating optimization (HBMO), and so on. On the other hand, there are other types of algorithms such as tabu search (TS) that has origins unrelated to natural processes. It is sometimes difficult to clearly assign an algorithm to one of these two classes (nature- and non-nature inspired), and many recently developed algorithms do not fit either class or may feature elements from both classes. Therefore, this classification is not particularly helpful. For instance, although TS is classified as a non-nature-inspired algorithm, it takes advantage of artificial intelligence aspects such as memory. It is therefore pertinent to argue whether or not the use of memory in the TS qualifies it as a nature-inspired algorithm.

2.4.2 Population-Based and Single-Point Search Algorithms

Some algorithms calculate iteratively one possible solution to an optimization problem. This means that the algorithms generate a single solution and they attempt to improve that solution in each iteration. Algorithms that work on a single solution are called trajectory methods and encompass local search-based meta-heuristics, such as TS. In contrast, population-based algorithms perform search processes that describe the evolution of a set of solutions in the search space. The GA is a good example of the population-based algorithms.

2.4.3 Memory-Based and Memory-Less Algorithms

A key feature of some meta-heuristic and evolutionary algorithms is that they resort to the search history to guide the future search for an optimal solution. Memory-less algorithms apply a Markov process to guide the search for a solution as the information they rely upon to determine the next action is the current state of the search process. There are several ways of using memory, which is nowadays recognized as one of the fundamental capabilities of advanced meta-heuristic and evolutionary algorithms.

2.5 Meta-Heuristic and Evolutionary Algorithms in Discrete or Continuous Domains

In meta-heuristic and evolutionary algorithms, each solution of an optimization problem is defined as an array of decision variables as follows:

$$X = \left(x_1, x_2, \ldots, x_i, \ldots, x_N\right) \tag{2.1}$$

where X = a solution of optimization problem, x_i = ith decision variable of the solution array X, and N = the number of decision variables.

Decision variables may be binary, discrete, or continuous values. Binary coding is used for Boolean decision variables of a binary nature (i.e., a situation occurs or it does not). Discrete values are used for problem with discrete decision space in which the decision variables are chosen from a predefined set of values. For instance, consider a two-dimensional problem with two decision variables x_1 and x_2 so that the values of x_1 and x_2 are chosen from the sets V_1 and V_2, respectively, where $V_1 = \{1.1, 4.5, 9.0, 10.25, 50.1\}$ and $V_2 = \{1, 7, 80, 100, 250\}$. Therefore, a feasible value of x_1 is 1.1, but 1.2 is not. In general, it can be stated that $x_i = \{v \mid v \in V_i\}$ as a condition defining the feasible values of the ith decision variable.

A class of discrete problems is that in which the decision variables must take integer values. A classic example is an optimization problem searching for the optimal numbers of individuals to be chosen from among K groups to make one consolidated optimal group. In this case, the optimal variables must take their values from the sets [0, 1, 2, ..., S_k], where S_k is the largest number of individuals in the kth group, $k=1, 2, ..., K$. Another example could be a problem with three-integer decision variables x_1, x_2, and x_3 whose allowable ranges are [0,6], [5,20], and [0,100], respectively. Therefore the feasible values of x_1 are {0,1,2,3,4,5,6}, and those of x_2 and x_3 are all the integer values between 5–20 and 0–100, respectively. In continuous problems the decision variables are real numbers contained between upper and lower boundaries and every value between boundaries is feasible. Constraints that involve functions of the decision variables, such as $2x_1+x_2+5x_3 \leq 20$, reduce the decision space. These constraints are commonly enforced by adding them as penalties to the objective function.

2.6 Generating Random Values of the Decision Variables

A few meta-heuristic and evolutionary algorithms are deterministic. Most of them, however, generate random values of the decision variables (possible solutions) at the start of the algorithm or during the search. There are algorithms that generate initial solutions deterministically, and during the search for an optimum, they generate random values of the decision variables. Decision variables are chosen randomly in the case of discrete domains. All the permissible values have an equal chance of being selected. Binary decision variables are randomly assigned the value zero or one with a probability equal to 0.5 each. Continuous decision variables are assigned values randomly between their lower and upper boundaries employing a suitable distribution function, such as the uniform distribution, a truncated normal distribution, or some other distribution. In most meta-heuristic and evolutionary algorithms, the uniform distribution is widely applied for random generation of decision variables.

2.7 Dealing with Constraints

Infeasible solutions occur in two ways. First, the values of the decision variables may be outside their allowable range. Second, even if all the decision variables are within their allowable range, they may be outside the feasible space, thus the solution is infeasible. Figure 2.5 shows one situation of feasible solution with feasible decision variables, one situation of an infeasible solution where the decision variables are within their ranges but outside the feasible space, and another case in which the solution is infeasible with the decision variables outside their ranges and outside the decision space.

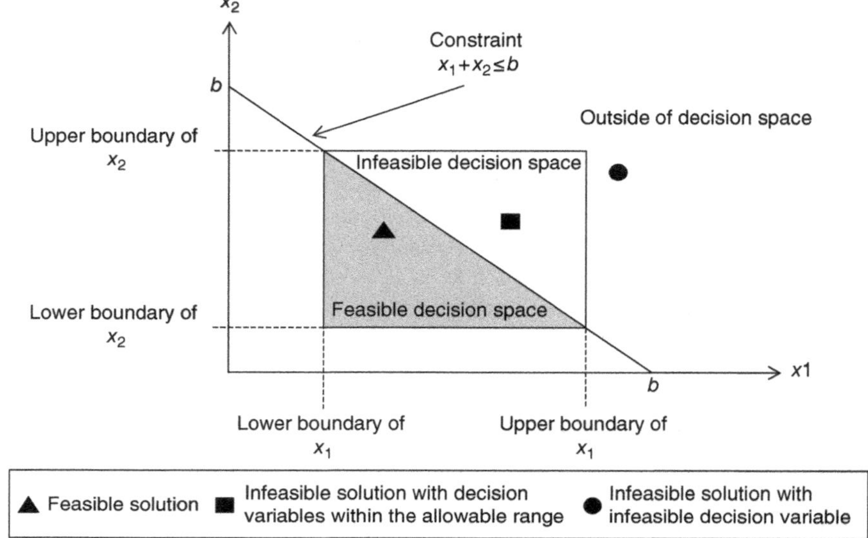

Figure 2.5 Different solutions in a two-dimensional decision space.

There are various methods to avoid infeasible decision variables, such as removing the infeasible space from the search, refinement, or using penalty functions that are discussed in the following text.

2.7.1 Removal Method

The removal method eliminates each possible solution that does not satisfy the constraints of the optimization model. Although the implementation of this method is simple, it has some disadvantages. First, this method does not distinguish between solutions with small and large constraints violations. Second, sometimes even infeasible solutions may yield clues about the optimal solution. For instance, although a solution might be infeasible, some of its decision variables may be the same as those of the optimum solution. Therefore, sometimes when a possible solution is deleted, its good features are eliminated from the search process.

2.7.2 Refinement Method

This method does not delete any of the infeasible solutions from the search process. Instead, the refinement method refines infeasible solutions to render them feasible solutions. For example, imagine that in building a structure two different materials A and B must be mixed in equal proportions and that variable B is a decision variable and variable A is a state variable. Assume that the sum of amounts A and B must equal 200 units. If the optimization problem chooses a B equal to 150, the constraints are not satisfied because an amount of A equal to 150 would violate the constraint on the sum of A and B. This situation calls for a refinement of the amount of B to satisfy both constraints, although optimality might be lost once the refinement is made. Refinement may produce an optimal solution or an infeasible solution. The refinement method uses features of an infeasible solution that might help in its refinement toward an optimal solution. The refinement method is cumbersome in that it is nontrivial to find ways to refine an infeasible solution and derive a feasible solution that is closer to an optimal one.

2.7.3 Penalty Functions

The application of penalty functions to avoid infeasible solutions overcomes the shortcomings of the removal and refinement methods. This method adds (or subtracts) a penalty function to the objective function of a minimization (or maximization) problem. The penalty function that is added or subtracted severely degrades the value of the objective function whenever a constraint is violated. Consider, for example, the following minimization problem:

$$Minimize\ f(X),\quad X=\left(x_1,x_2,\ldots,x_i,\ldots,x_N\right) \tag{2.2}$$

Subject to

$$G(X)>\delta_1 \tag{2.3}$$

$$H(X)<\delta_2 \tag{2.4}$$

$$Z(X)=\delta_3 \tag{2.5}$$

The penalized objective function, or fitness function, for this minimization problem is achieved by adding penalties to the objective function as follows:

$$F(X)=f(X)+Penalty \tag{2.6}$$

where

$$Penalty=\left(\vartheta(G(X))\times\phi_1\right)+\left(\xi(H(X))\times\phi_2\right)+\left(\psi(z(x))\times\phi_3\right) \tag{2.7}$$

in which the penalty on a violation of the constraint $G(X)$:

$$\vartheta\left(G(X)\right)=\begin{cases}0 & if\ G(X)>\delta_1\\ 0 & if\ G(X)\le\delta_1\end{cases} \tag{2.8}$$

$$\phi_1=\alpha_1\times\left(\delta_1-G(X)\right)^{\beta_1}+C_1 \tag{2.9}$$

The penalty on a violation of the constraint $H(X)$:

$$\xi(H(X)) = \begin{cases} 0 & \text{if } H(X) < \delta_2 \\ 0 & \text{if } H(X) \geq \delta_2 \end{cases} \tag{2.10}$$

$$\phi_2 = \alpha_2 \times \left(H(X) - \delta_2\right)^{\beta_2} + C_2 \tag{2.11}$$

The penalty on a violation of the constraint $Z(X)$:

$$\psi(Z(X)) = \begin{cases} 0 & \text{if } Z(X) = \delta_3 \\ 0 & \text{if Otherwise} \end{cases} \tag{2.12}$$

$$\phi_3 = \alpha_3 \times \left(\left|Z(X) - \delta_3\right|\right)^{\beta_3} + C_3 \tag{2.13}$$

where X = solution of the optimization problem, $f(X)$ = value of the objective function of solution X, $G(X)$ = a constraint whose value exceeds δ_1, $H(X)$ = a constraint whose value is less than δ_2, $Z(X)$ = a constraint whose value equals δ_3, $F(X)$ = penalized objective function, *Penalty* = total value of the penalty on constraints violations, ϕ_1 = penalty for constraint $G(X)$, ϕ_2 = penalty for constraint $H(X)$, ϕ_3 = penalty for constraint $Z(X)$, and α_k, β_k, and C_k (k = 1, 2, 3) = constant values for adjusting the magnitude of the penalty function. In a maximization problem the penalized objective function, or fitness function, is written by subtracting penalties from the objective function as follows:

$$F(X) = f(X) - Penalty \tag{2.14}$$

The coefficients α_k and β_k quantify the magnitude of constraints violations. For example, for $\beta_k = 1$ the amount of the penalty increases linearly with the increase in constraint violation; α_k determines the slope of the penalty function. C_k changes the value of the fitness of an infeasible solution independently of the magnitude of the constraint violation. The user-specified values of α_k, β_k, and C_k impact the performances of the penalty function and of the optimization algorithm searching for an optimal solution. In fact, employing penalty functions modifies the mapping between the objective function and the decision space. It is possible that the optimal solution of the unconstrained penalized objective function might differ from the optimal solution of the constrained objective function.

The specification of proper values for α_k, β_k, and C_k relies on experience with specific types of optimization problems and on reliance on sensitivity analysis. In the latter approach the analyst tries several combinations of α_k, β_k, and C_k and applies the optimization algorithm to calculate solutions and compare them. The combination of the penalty coefficients that yields the best solutions becomes the best choice of penalty coefficients.

2.8 Fitness Function

The penalized objective function is called the fitness function. Therefore, the fitness function is written as follows:

$$F(X) = f(X) \pm Penalty \tag{2.15}$$

where X = solution of the optimization problem, $f(X)$ = objective function of solution X, and $F(X)$ = fitness function (penalized objective function) of solution X. The penalty is added (or subtracted) in a minimization (maximization) problem.

2.9 Selection of Solutions in Each Iteration

Selection refers to choosing some solutions from a set of solutions during the algorithmic calculations. In some meta-heuristic and evolutionary algorithms, not all current solutions are employed to generate new solutions. The selection operators bypass many current solutions. The selection of solutions among the current set of solution is done randomly or deterministically based on the algorithm. In some algorithms, although all the current solutions are used to generate new solutions, not all the new solutions are accepted. Only those that have relatively high merit are applied in the search process. Selection of some newly generated solutions among all the generated new solutions can be done randomly or deterministically. Usually such selection is done based on the fitness values of the decision variables, which are the current solutions in any algorithmic iteration. This means that in random selection methods, a higher probability of selection is assigned to superior solutions over inferior solutions. In deterministic selection methods usually the best solution(s) is (are) selected from a set of solutions. The selection of current solutions to generate new solutions in the algorithmic iterations has an important role in finding the optimum. Therefore, the selective pressure is an important factor in meta-heuristic and evolutionary algorithms. A selection method with high selective pressure most likely selects the best solutions and eliminates the worst ones at every step of the search. In contrast, a selection method with a very low selective pressure ignores the fitness values of the current solutions and assigns the same probability of selection to the diverse solutions featuring different fitness values. Figure 2.6 shows a set of solutions of an imaginary maximization problem so that the solutions are ranked based on their fitness values, and an example probability of selection is prescribed to each solution, employing a high selective pressure or a low selective pressure.

It is seen in Figure 2.6 that a low selective pressure assigns nearly equal probability of selection to diverse solutions regardless of their fitness. A high

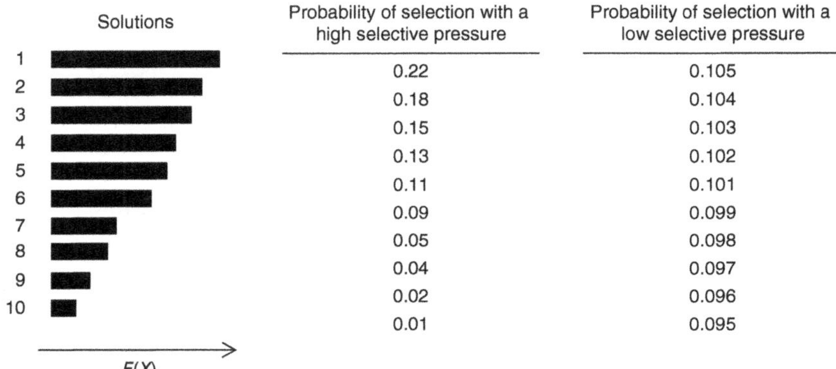

Solutions	Probability of selection with a high selective pressure	Probability of selection with a low selective pressure
1	0.22	0.105
2	0.18	0.104
3	0.15	0.103
4	0.13	0.102
5	0.11	0.101
6	0.09	0.099
7	0.05	0.098
8	0.04	0.097
9	0.02	0.096
10	0.01	0.095

Figure 2.6 Selection probability of a set of solutions 1–10 of a hypothetical maximization problem.

selective pressure, on the other hand, assigns higher probability of selection to the fittest (largest) solutions. Recall that the uniform distribution assigns an equal probability of selection to all the solutions in each iteration. A selection process that chooses solutions based on the uniform distribution does not apply selective pressure. Unlike deterministic selection and selection with uniform distribution, there are other selection methods that allow the user to adjust the selective pressure or the methods themselves change the selective pressure automatically. In fact, one of the differences between meta-heuristic and evolutionary algorithms is how they select solutions. Several algorithms do not implement a selection process, whereas others do. Among common selection methods are the Boltzmann selection, the roulette wheel, the tournament selection, and others. These selection methods are described in this book.

2.10 Generating New Solutions

A key step of the meta-heuristic and evolutionary algorithms is generating new solution(s) from the current one in each iteration. Each iteration of an algorithm is completed by generating new solution(s). Each algorithm generates new solutions differently from others. In all cases, however, all the algorithms rely on the current solutions to generate new ones. In fact, new solutions are usually in the neighborhood of a previous solution, they are a combination of two or more old solutions or they are randomly generated solutions whose acceptance for entering the search process is determined by comparison with previous solutions. The methods employed by leading meta-heuristic and evolutionary algorithms to generate new solutions iteratively are described in this book.

2.11 The Best Solution in Each Algorithmic Iteration

In some algorithms the best solution in each iteration is highlighted. Some algorithms keep the best solution in an iteration without any changes and carry it to the next iteration until a better solution is generated, at which time the better solution takes its place. Other algorithms such as the HBMO preserve the best solution in each iteration and assign it a larger weight to generate new solutions. Each algorithm names the best solution of an iteration differently, such as "base point" in the pattern search (PS) or "the queen" in the HBMO algorithm and so on.

2.12 Termination Criteria

Each iteration of an algorithm finishes with the generation of new solutions. The algorithm evaluates the fitness function of each solution and moves on to the next iteration, or it is terminated if the termination criteria are satisfied. Three prevalent termination criteria are the number of iterations, a threshold of improvement of the fitness function in consecutive iterations, and the run time.

The first criterion sets a number of iterations so that the algorithm continues for predefined number of iterations. For example, the maximum number of algorithmic iterations may be set at 10^6, at which time it will stop. The main disadvantage with this criterion is that the analyst does not know a priori how many iterations are good enough. Thus, the algorithm might be stopped prematurely when the current solution is far from optimal or it could reach a near-optimal solution quickly and thereafter continue replicating that solution without further improvement and inflicting unnecessary computational burden.

The second criterion stops the execution of the algorithm whenever the difference between the solutions pertaining to two or more consecutive iterations falls below a user-specified threshold. A disadvantage of this method is that the solution achieved may be a local optimum. The meta-heuristic and evolutionary algorithms usually employ randomness or other tools to escape from local solutions if they are allowed to keep searching even if a threshold is met between a few consecutive iterations.

The maximum run time criterion stops the algorithm after a specified processing time is complete and reports the best solution achieved up to that time without consideration to the number of iterations performed or the rate of improvement of the solution. The main limitation of this criterion is identical to that of which defines a maximum number of iterations, that is, it is generally unknown how much time is necessary to reach a near-optimal solution.

There are other termination or stopping criteria that rely on special features of a search algorithm. These are discussed in various parts of this book.

2.13 General Algorithm

This section presents a general algorithm that encompasses most or all of the steps found in meta-heuristic and evolutionary algorithms. This allows the comparison of specific algorithms covered in this book to the general algorithm. All meta-heuristic and evolutionary algorithms begin by generating initial (possible or tentative) solution(s), which is (are) named "old" solutions and is (are) improved by the algorithm. The search algorithm iteratively improves "old" or known solutions with improved "new" ones until a termination criterion is reached. The steps of a general algorithm are as follows:

Step 0: Read input data.
Step 1: Generate initial possible or tentative solutions randomly or deterministically.
Step 2: Evaluate the fitness values of all current solutions.
Step 3: Rename the current solutions as old solutions.
Step 4: Rank all the old solutions and identify the best among them, those with relatively high fitness values.
Step 5: Select a subset of the old solutions with relatively high fitness values.
Step 6: Generate new solutions.
Step 7: Evaluate the fitness value of the newly generated solutions.
Step 8: If termination criteria are not satisfied, go to step 3; otherwise go to step 9.
Step 9: Report all the most recently calculated solutions or the best solution achieved at the time when the algorithm terminates execution.

Figure 2.7 illustrates the flowchart of the general algorithm.

2.14 Performance Evaluation of Meta-Heuristic and Evolutionary Algorithms

An evolutionary or meta-heuristic algorithm starts with initial solutions and attempts to improve them. Figure 2.8 depicts the progress of an algorithm that gradually convergences to a near optimum of imaginary hypothetical minimization problem. The convergence of an algorithm may be traced by graphing or monitoring the fitness value of the best solution against the number of iterations or the run time. In addition to the number of iterations and the run time, another variable called the number of functional evaluations (NFE) may also be employed for tracing the convergence of an algorithm. The NFE equal the number of evaluations of the fitness function executed during the application of a search algorithm and equal the number of initial solutions plus the product of the number of iterations performed during algorithmic execution multiplied by the number of solutions generated in each iteration of the algorithm.

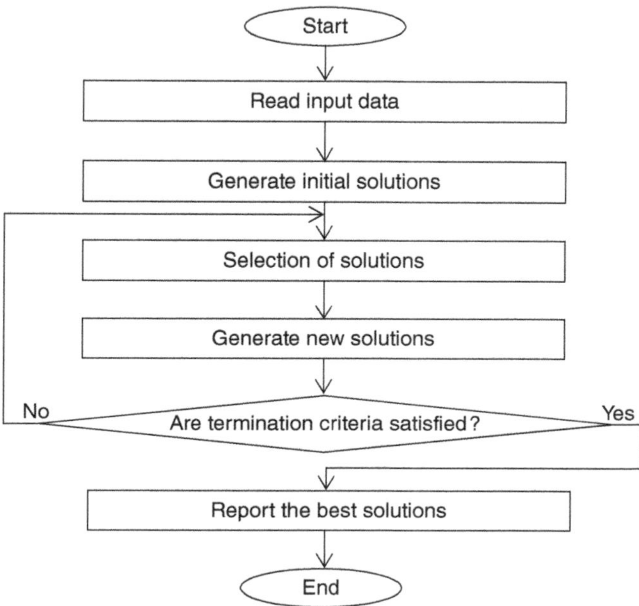

Figure 2.7 The flowchart of the general algorithm.

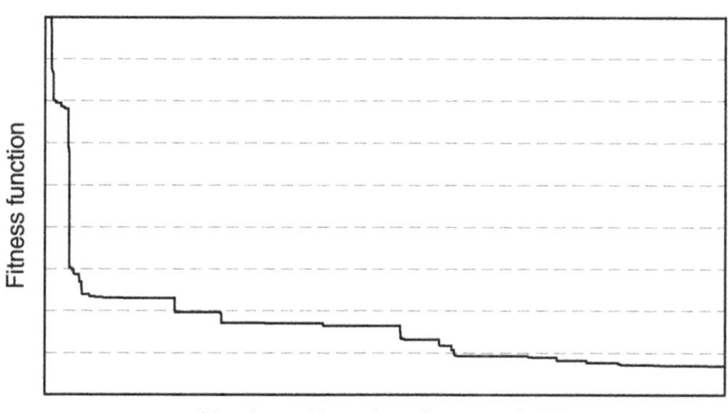

Figure 2.8 Convergence history of an optimization algorithm toward the best solution in a minimization problem.

It is seen in Figure 2.8 how the value of the fitness function improves rapidly in the first iterations and eventually convergences to the optimal solution after a relatively large number of iterations. Furthermore, the fitness function improves or is maintained at the same level from one iteration to the next. Figure 2.8 is typical of the PS or HBMO algorithm, which improves or keeps the best solution calculated

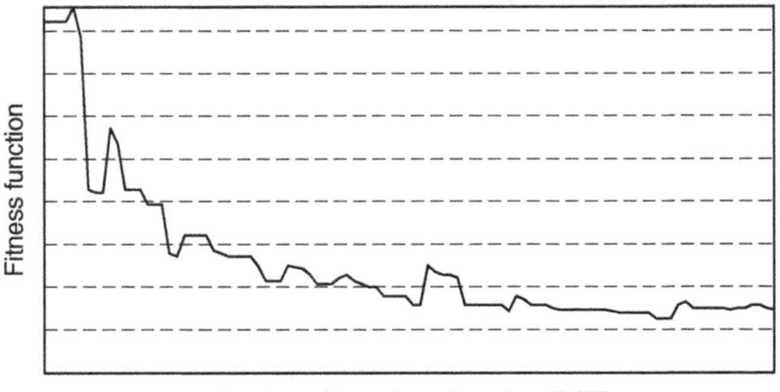

Figure 2.9 Convergence of an optimization algorithm in which the best solution is not always transferred to the next iteration during the search in a minimization problem.

in the current iteration and use it in the next iteration. There are other algorithms in which the best solution(s) is not transferred to the next iteration, the standard GA being a case in point. The convergence of the latter type of algorithms resembles the chart depicted in Figure 2.9, in which the best solution of an iteration may be worse than that of the previous iteration, even though, overall, there is a convergence toward an optimal solution as the number of iterations increases.

The NFE seem to be the best variable to measure the speed with which an algorithm converges to an optimal solution. The run time, on the other hand, may be affected by programming skill, type of programming language, computer speed, and other factors that are not algorithm dependent.

Another consideration is that most meta-heuristic and evolutionary algorithms apply a random search and few of them are deterministic algorithms. Random-search algorithms require several runs to solving a given problem. Each run most likely produces a slightly different near-optimal solution. Therefore, in judging the performance of a random-search algorithm, several runs of the algorithm are performed when solving a given problem. The following factors determine the quality of the algorithm's performance: (1) capacity to reach near-optimal solutions consistently, that is, across several runs solving a given problem, and (2) speed of the solution algorithm in reaching near-optimal solutions. Algorithmic reliability is defined based on the variance of the solutions' fitness-function values achieved in several runs of an algorithm reported as final solutions by different runs of an algorithm. A reliable algorithm is one that converges to very similar near-optimal solutions in roughly the same NFE. A reliable algorithm features relatively small variance of the solutions' fitness-functions values. Figure 2.10 demonstrates this feature for three different runs of a hypothetical algorithm.

It is seen in Figure 2.10 that the various runs start with different, randomly generated initial solutions and converge to the same near-optimal solution in about the same NFE, a defining feature of a reliable algorithm. Figure 2.11

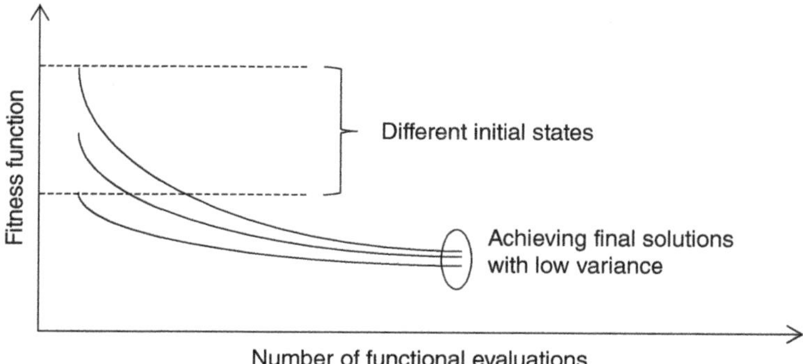

Figure 2.10 Convergence of different runs of an optimization algorithm toward near-optimal solutions of a minimization problem.

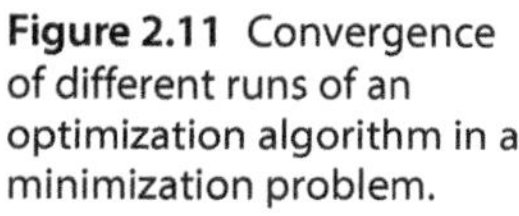

Figure 2.11 Convergence of different runs of an optimization algorithm in a minimization problem.

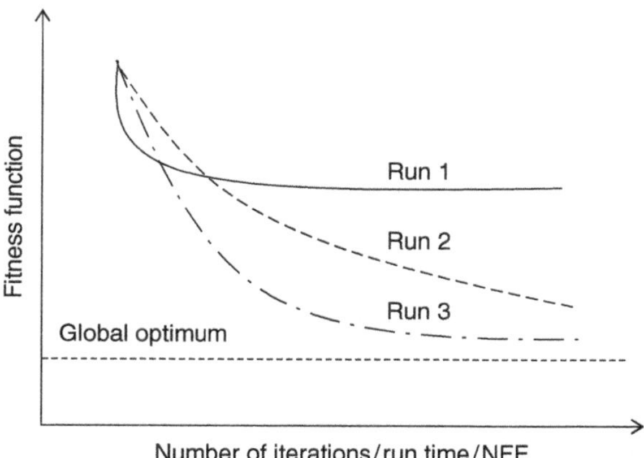

depicts different runs of an algorithm in a minimization problem where all runs start from the same state. Run No. 1 converges fast but to a solution that is clearly nonoptimal. Run No. 2 reaches an acceptable solution that is close enough to the global optimum, but its convergence speed is relatively low. Run No. 3 achieves a near-optimal solution with the fastest convergence rate.

Table 2.6 provides a summary of the solution performance of an algorithm in k runs executed to solve a minimization problem.

2.15 Search Strategies

The need for solving various problems with their peculiar decision spaces led to the development of algorithms inspired by natural phenomena or that mimic human intelligence. Some critics argued that a few of the newly

Table 2.6 Recommended reporting of the solutions calculated in *k* runs of an algorithm that solves a minimization problem.

Criteria	Result
Run 1	F_1
Run 2	F_2
⋮	⋮
Run *k*	F_k
Best solution in *k* runs	$F_r, F_r \le F_i, \quad i = 1,2,\ldots,K$
Average solution in *k* runs	$\frac{\sum_{r=1}^{k} F_r}{k}$
Worst solution in *k* runs	$F_r, \ F_r \ge F_i, \quad i = 1,2,\ldots,K$
Standard deviation of solutions in *k* runs	$\frac{\sqrt{\sum_{r=1}^{k} (F_r - Average)^2}}{k}$
Coefficient of variation of solutions in *k* runs	$\frac{st.deviation}{Average}$

F_k = the fitness value of the solution achieved with the *k*th run of the optimization algorithm; *k* = the total number of independent runs of the optimization algorithm; *Average* = the average solution in *k* runs; *st.devision* = standard deviation of solutions in *k* runs.

developed algorithms, although inspired by different phenomena, in practice, are a repetition of previously developed algorithms (e.g., Sörensen, 2013). In contrast, other researchers proved the differences between newly developed algorithms and old ones (e.g., Saka et al., 2016). All meta-heuristic and evolutionary algorithms have common features present in the general algorithm introduced this chapter. There are, however, significant differences among them such as in generating the initial and new solutions and in selecting new solutions. Each algorithm searches the decision space differently, and their efficiencies in solving specific problems vary. For example, algorithms that use a high selective pressure in their selection stage and emphasize on searching near the best found solutions can search single-modal decision spaces efficiently. However, their performance decreases searching multimodal decision spaces where there are several local optima because of the risk of being entrapped in local optima. On the other hand, the algorithms apply selection methods with low selective pressure and search the decision space randomly, thus implementing a thorough search all over the decision space and reducing the risk of entrapment near local optima, which makes them more effective in solving problems with multimodal decision spaces. However, these algorithms make a large number of calculations in solving problems with single-modal decision space where there is no risk of entrapment near a local optimum. Also, most meta-heuristic and evolutionary algorithms have parameters that regulate their performance. The choice of these parameters affects their search strategies. Other algorithms set their parameters automatically. For these reasons knowing the principles on which each algorithm works is essential for users applying them, who must choose an appropriate algorithm to solve specific optimization problems. Twenty algorithms are described in the remainder of this book and are the leading meta-heuristic and evolutionary algorithms known to date: PS, GA, simulated annealing (SA), TS, ACO, particle swarm optimization (PSO), differential evolution (DE), harmony search (HS), shuffled frog-leaping algorithm (SFLA), HBMO, invasive weed optimization (IWO), central force optimization (CFO), biogeography-based optimization (BBO), firefly algorithm (FA), gravity search algorithm (GSA), bat algorithm (BA), plant propagation algorithm (PPA), water cycle algorithm (WCA), symbiotic organisms search (SOS), and comprehensive evolutionary algorithm (CEA).

2.16 Conclusion

This chapter provided an introduction to meta-heuristic and evolutionary algorithms by explaining different methods employed for searching the decision space. The components of meta-heuristic and evolutionary algorithms were highlighted, and the features of several engineering optimization problems were explained. Other related topics were introduced in this chapter, among which are coding meta-heuristic and evolutionary algorithms, dealing with constraints, selection of solutions, and so on. A general algorithm was presented that includes the most common traits of meta-heuristic and evolutionary algorithms. This general algorithm shall serve as a useful baseline for comparing various algorithms. Lastly, this chapter explained methods for evaluating the performance of meta-heuristic and evolutionary algorithms.

References

Saka, M. P., Hasançebi, O., and Geem, Z. W. (2016). "Metaheuristics in structural optimization and discussions on harmony search algorithm." Swarm and Evolutionary Computation, 28, 88–97.

Sörensen, K. (2013). "Metaheuristics: The metaphor exposed." International Transaction in Operational Research, 22(1), 3–18.

3

Pattern Search

Summary

This chapter explains the pattern search (PS) algorithm. The PS is a meta-heuristic algorithm that is classified as a direct search method.

3.1 Introduction

Hooke and Jeeves (1961) called the pattern search (PS) a family of numerical optimization methods that do not require calculating the gradient of the objective function in solving optimization problems. Tung (1984) employed the PS algorithm to calibrate the Muskingum model. Neelakantan and Pundarikanthan (1999) calculated an optimal hedging rule for water supply reservoir systems. They applied a neural network model to speed up the optimization process without considering the number of functional evaluations needed in the simulation of the reservoir system operation. Al-Sumait et al. (2007) presented a new method based on the PS algorithm to solve a well-known power system economic load dispatch (ELD) problem with valve-point effect. Bozorg-Haddad et al. (2013) implemented the PS for groundwater model calibration and compared the performance of the PS with that of the particle swarm optimization (PSO) algorithm. Groundwater models are computer models that simulate and predict aquifer conditions in response to groundwater withdrawal or recharge or some other stress on an aquifer. Mahapatra et al. (2014) proposed a hybrid firefly algorithm and pattern search (h-FAPS) technique for a static synchronous series compensator (SSSC)-based power oscillation damping controller design. Khorsandi et al. (2014) applied optimization techniques including the genetic algorithm (GA) and PS for the identification of the location and quantity of surface-water pollutants.

Meta-Heuristic and Evolutionary Algorithms for Engineering Optimization,
First Edition. Omid Bozorg-Haddad, Mohammad Solgi, and Hugo A. Loáiciga.

3.2 Pattern Search (PS) Fundamentals

The PS is a direct search method. Direct search methods solve a variety of numerical problems with emphasis on the use of simple strategies that make them better suited for implementation in modern computers than classical methods (e.g., linear of nonlinear programming). The qualifier "direct search" refers to sequential examination of trial solutions. Direct search compares each trial solution with the best solution previously obtained, and the result of the comparison determines what the next trial solution will be. Direct search techniques employ straightforward search strategies. These techniques have features that distinguish them from classical methods and have solved problems that classical methods could not solve. Also, direct search techniques converge faster to the solutions of some problems than classical methods. Direct search techniques rely on repeated identical arithmetic operations with simple logic that are easily coded for computer calculations. Direct search techniques converge to near-optimal solutions, which is also a feature of meta-heuristic and evolutionary algorithms (Hooke and Jeeves, 1961).

Direct search methods randomly select a point B and call it a base point. A second point, P1, is randomly selected, and if it is better than B, it replaces the base point; if not, B remains the base point. This process continues with each new randomly selected point being compared with the current base point. The "strategy" for selecting new trial points is determined by a set of "states" that constitutes the memory of the algorithm. The number of states is finite. There is an arbitrary initial state and a final state that stops the search. The other states represent various situations that arise as a function of the results of the trials made. The kind of strategy implemented to select new points is dictated by various aspects of the problem, including the structure of the decision space of the problem. The strategy includes the choice of the initial base point, the rules of transition between states, and the rules for selecting trial points as a function of the current state and the base point. Direct search designates a trial point as a move or step from the base point. The move is a success if the trial point is better than the base point, or it is a failure otherwise. The states make up part of the logic, influencing moves to be proposed in the same general direction as those that have recently succeeded. The states suggest new directions if recent moves have failed. The states decide when no further progress can be made. The fact that no further progress can be made does not always indicate that the solution has been found.

Solutions of the optimization problem calculated by the PS algorithm are points in an N-dimensional space with N denoting the number of decision variables. Trial points refer to new solutions. The process of going from a given point to a trial point is called a move. A move may be successful if the trial

Table 3.1 The characteristics of the PS.

General algorithm (see Section 2.13)	Pattern search
Decision variable	Coordinates of point's position
Solution	Point
Old solution	Base point
New solution	Trial point
Best solution	Base point
Fitness function	Desirability of the base point
Initial solution	Random point
Selection	Comparison of the trial point with base point
Generate new solution	Exploratory and pattern move

point is better than the base point; otherwise it is a failure. The PS finds the correct route to achieve optima from analyzing the failures and success of trial points. In other words, failure or success of trial points affects the direction and length of the steps of the movements in next stages. Table 3.1 lists the characteristics of the PS algorithm.

The PS generates a sequence of solutions that produces a mesh around the first solution and approaches an optimal solution. An initial solution is randomly generated and known as the base point. Next, trial points (new solutions) are generated. There are two patterns to generate trial points. The first pattern is an exploratory move designed to acquire knowledge about the decision space. This knowledge issues from the success or failure of the exploratory moves without regard to any quantitative appraisal of the values of the fitness functions. It means that exploratory moves determine a probable direction for a successful move. The second one is a pattern move that is designed to use the information achieved by the exploratory moves to find the optimal solution. After generating new solutions with the exploratory moves, the PS algorithm computes the fitness function at the mesh points and selects one whose value is better than the first solution's fitness value. The searching point is transferred to the new solution if there is a point (solution) with a better obtained fitness function among the generated solutions. At this stage, the expansion coefficient is applied to generate new solutions. Thus, the new mesh size is larger than the previous one. In contrast, if there is no better solution in the generated solutions, the contraction coefficient is applied and the mesh size is limited to a smaller value. Figure 3.1 shows the flowchart of the PS algorithm.

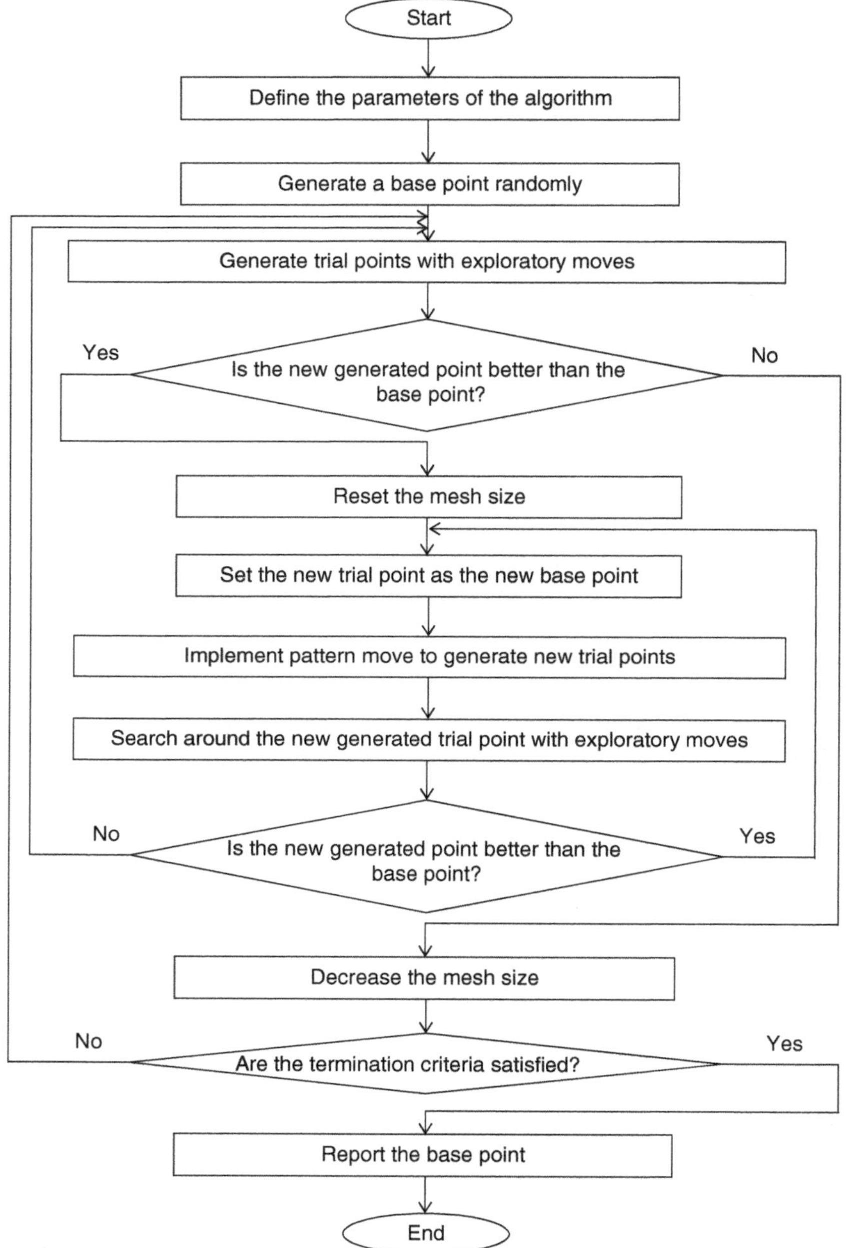

Figure 3.1 The flowchart of the PS.

3.3 Generating an Initial Solution

Each possible solution of the optimization problem calculated by the PS is a point in the decision space. Therefore, in an N-dimensional optimization problem, the position of the point in that space is a decision variable of the optimization problem that constitutes an array of size $1 \times N$. The PS starts with a single solution that is written as a matrix or row vector of size $1 \times N$ as follows:

$$Point = X = \left(x_1, x_2, \ldots, x_i, \ldots, x_N\right) \tag{3.1}$$

where X = a solution of the optimization problem; x_i = ith decision variable of the solution X; and N = number of decision variables. The decision variable values $(x_1, x_2, x_3, \ldots, x_N)$ are defined appropriately for continuous and discrete problems. The PS algorithm starts with an initial possible or tentative solution that is randomly generated (see Section 2.6) and known as the base point. Subsequently, trial points (solutions) are generated around the base point.

3.4 Generating Trial Solutions

Trial solutions are new solutions of the optimization problem that may potentially be the next base point. There are two patterns with which to generate trial solutions. The first pattern is an exploratory move designed to acquire knowledge about the decision space. The second one is a pattern move that is designed to use the information achieved by the exploratory moves to find the optimal solution.

3.4.1 Exploratory Move

The exploratory move obtains information about the decision space of the optimization problem. This knowledge is derived from the success or failure of the exploratory moves without regard to any quantitative appraisal of the values of the fitness functions. In each exploratory move the value of a single coordinate is changed, and the effect of this change on the fitness value is evaluated. The fitness value after the move is evaluated, and the move is successful if the newly calculated fitness value improves the fitness value prevailing before the move. Otherwise, the move is a failure. The purpose of the exploratory move is to find a direction of improvement. This purpose is achieved by perturbing the current point by small amounts in each of the variable directions and determining if the fitness function value improves or worsens. There are two patterns to generate solutions by an exploratory move including (1) generalized pattern search (GPS) and (2) mesh adaptive direct search (MADS). The number of solutions generated by GPS about the first solution equals $2N$, which are produced as follows, where the solution X is a matrix of size $1 \times N$:

$$X_1^{(new)} = \mu \cdot \begin{bmatrix} 1 & 0 & 0 & \dots & 0 \end{bmatrix}_{1\times N} + X \tag{3.2}$$

$$X_2^{(new)} = \mu \cdot \begin{bmatrix} 0 & 1 & 0 & \dots & 0 \end{bmatrix}_{1\times N} + X \tag{3.3}$$

$$\vdots$$

$$X_N^{(new)} = \mu \cdot \begin{bmatrix} 0 & 0 & \dots & 0 & 1 \end{bmatrix}_{1\times N} + X \tag{3.4}$$

$$X_{N+1}^{(new)} = \mu \cdot \begin{bmatrix} -1 & 0 & 0 & \dots & 0 \end{bmatrix}_{1\times N} + X \tag{3.5}$$

$$X_{N+2}^{(new)} = \mu \cdot \begin{bmatrix} 0 & -1 & 0 & \dots & 0 \end{bmatrix}_{1\times N} + X \tag{3.6}$$

$$\vdots$$

$$X_{2N}^{(new)} = \mu \cdot \begin{bmatrix} 0 & 0 & \dots & 0 & -1 \end{bmatrix}_{1\times N} + X \tag{3.7}$$

in which μ = mesh size; X = base point; and $X^{(new)}$ = trial solution.

MADS generates $N+1$ new solutions about the base point (X) as follows:

$$X_1^{(new)} = \mu \cdot \begin{bmatrix} 1 & 0 & 0 & \dots & 0 \end{bmatrix}_{1\times N} + X \tag{3.8}$$

$$X_2^{(new)} = \mu \cdot \begin{bmatrix} 0 & 1 & 0 & \dots & 0 \end{bmatrix}_{1\times N} + X \tag{3.9}$$

$$\vdots$$

$$X_N^{(new)} = \mu \cdot \begin{bmatrix} 0 & 0 & \dots & 0 & 1 \end{bmatrix}_{1\times N} + X \tag{3.10}$$

$$X_{N+1}^{(new)} = \mu \cdot \begin{bmatrix} -1 & -1 & -1 & \dots & -1 \end{bmatrix}_{1\times N} + X \tag{3.11}$$

Figure 3.2 displays trial solutions generated by GPS and MADS methods in a two-dimensional decision space.

The PS algorithm computes the fitness function of all new solutions after generating new points with either GPS or MADS and further evaluates the one with the best fitness value among them. A comparison is made between the best new solution and the base point. If there is a solution with a better fitness value in the generated solutions, the search point is transferred to this new point. At this stage, the expansion coefficient is used to generate new solutions. Thus, the new mesh size is larger than the previous one. In contrast, if there is no better solution in the generated solutions, the contraction coefficient is

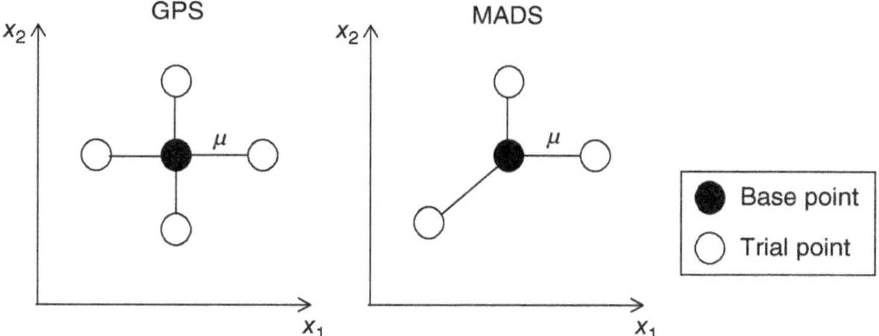

Figure 3.2 Meshes generated by the GPS and MADS methods in a two-dimensional decision space.

applied and the mesh size is reduced to a smaller value. After a successful exploratory move, the pattern move is implemented to further the search in a direction of likely improvement.

3.4.2 Pattern Move

The pattern move is designed to use the knowledge acquired in the exploratory moves. When the exploratory move achieves a better point, the new generated point becomes the new base point. A new trial point is generated based on the previous point and the current base point is calculated as follows:

$$X^{(new)} = X' + \alpha \cdot (X - X') \tag{3.12}$$

in which $X^{(new)}$ = the new trial solution; X' = the previous base point; X = the current base point; and α = a positive acceleration factor.

The pattern move from a given base point repeats the combined moves from the previous base point. The reasoning for this type of movement is the presumption that whatever constituted a successful set of moves in the past is likely to prove successful again. Each pattern move is immediately followed by a sequence of exploratory moves that continually revise the pattern and could improve the new trial solution. If the new generated trial solution is better than the current base point, it becomes the new base point and the pattern move is implemented again to generate new trial point. Otherwise the current base point is not changed, and only exploratory moves are implemented about the base point (see Figure 3.1).

A pattern move first obtains a tentative trial solution and then finds a trial solution with an exploratory search. Pattern moves are repeated as long as they are successful (i.e., improve the fitness function) and usually become longer and longer steps. As soon as a pattern move fails, the pattern move is retracted and the algorithm goes back to an exploratory search about the best point so far calculated.

3.5 Updating the Mesh Size

If there is no better solution among the generated solutions by the exploratory move, the contraction coefficient is applied and the mesh size (μ) is reduced to a smaller value. For any given value of the mesh size, the algorithm reaches an impasse when all the exploratory moves from the base point fail. In this situation it is necessary to reduce the mesh size (μ) to continue the search. The magnitude of the reduction is sufficient to permit a new pattern to be established. However, too large a reduction in the mesh size slows the search. For this reason when all trial points are worse than the base point, the mesh size is decreased as follows:

$$\mu^{(new)} = \mu - \delta \tag{3.13}$$

in which $\mu^{(new)}$ = new mesh size and δ = decreasing step of mesh size, which is a user-defined parameter of the algorithm.

If there is a point (solution) with a better fitness value among the generated solutions, the search point is transferred to this new point. At this stage, the expansion coefficient is applied to generate new solutions. Thus, the new mesh size is larger than the previous one. Whenever the exploratory move reaches a better point, the mesh size (μ) is reset to the initial value as follows:

$$\mu^{(new)} = \mu_0 \tag{3.14}$$

in which μ_0 = the initial value of μ, which is determined by the analyst as a parameter of the algorithm.

3.6 Termination Criteria

The termination criterion determines when to stop the algorithm. Selecting a good termination criterion is essential to avoid a premature stoppage whereby a suboptimal or a local optimal solution is calculated. On the other hand, it is desirable to avoid unnecessary calculations beyond a solution that cannot be improved once it is reached. A common termination criterion for the PS is the mesh size. The final termination of the search is made when the mesh size is sufficiently small to insure that the optimum has been closely approximated. Limiting the number of iterations, or the run time, and monitoring the improvement of the solution in consecutive iterations are other termination criteria that can be implemented with the PS algorithm.

3.7 User-Defined Parameters of the PS

The value of acceleration factor (α), initial mesh size (μ_0), decreasing step of the mesh size (δ), and termination criterion are user-defined parameters of the PS. A good choice of the parameters is dependent of the decision space of a particular problem, and usually the optimal parameter setting for one problem is of limited utility for other problems. A reasonable method for finding appropriate values of the algorithmic parameters is performing sensitivity analysis. This consists of experimenting with multiple combinations of parameters with which the algorithm is run. The results from the various combinations are compared, and the analyst chooses the parameter set that yields the best optimization results.

3.8 Pseudocode of the PS

```
Begin
  Input parameters of the algorithm and initial data
  Let X' = previous base point; X^(new) = new generated
    point and X = the current base point
  Generate a base point (X) randomly and calculate its
    fitness function
  While (the termination criteria are not satisfied)
     Generate new points about X by exploratory moves
       and evaluate their fitness values
     Set X^(new) = the best new generated point
     If X^(new) is better than X
         Reset the mesh size
         While (X^(new) is better than X)
             Set X' = X
             Set X = X^(new)
             Obtain a new X^(new) by pattern move with X' and X
             Generate new points around X^(new) by exploratory
               moves and evaluate their fitness functions
             Set X^(new) = the best new generated point
         End while
     Otherwise
         Decrease the mesh size
     End if
  End while
  Report the solution
End
```

3.9 Conclusion

The PS is a meta-heuristic algorithm of the direct search type. This chapter explained the workings of the PS as a direct search method and provides pseudocode of the algorithm.

References

Al-Sumait, J. S., AL-Othman, A. K., and Sykulski, J. K. (2007). "Application of pattern search method to power system valve-point economic load dispatch." International Journal of Electrical Power & Energy Systems, 29(10), 720–730.

Bozorg-Haddad, O., Tabari, M. M. R., Fallah-Mehdipour, E., and Mariño, M. A. (2013). "Groundwater model calibration by meta-heuristic algorithms." Water Resources Management, 27(7), 2515–2529.

Hooke, R. and Jeeves, T. A. (1961). "Direct search solution of numerical and statistical problems." Journal of the ACM, 8(2), 212–229.

Khorsandi, M., Bozorg-Haddad, O., and Mariño, M. A. (2014). "Application of data-driven and optimization methods in identification of location and quantity of pollutants." Journal of Hazardous, Toxic, and Radioactive Waste, 19(2), 04014031.

Mahapatra, S., Panda, S., and Swain, S. C. (2014). "A hybrid firefly algorithm and pattern search technique for SSSC based power oscillation damping controller design." Ain Shams Engineering Journal, 5(4), 1177–1188.

Neelakantan, T. R. and Pundarikanthan, N. V. (1999). "Hedging rule optimisation for water supply reservoirs system." Water Resources Management, 13(6), 409–426.

Tung, Y. K. (1984). "River flood routing by nonlinear Muskingum method." Journal of Hydraulic Engineering, 111(12), 1447–1460.

4

Genetic Algorithm

Summary

This chapter describes the genetic algorithm (GA), which is a well-known evolutionary algorithm. First, a brief literature review of the GA is presented, followed by a description of the natural process that inspires the algorithm and how it is mapped to the GA. The steps of the standard GA are described in depth. A pseudocode of the GA closes this chapter.

4.1 Introduction

One of the best-known evolutionary algorithms is the genetic algorithm (GA) developed by Holland (1975) and popularized by Goldberg (1989). There are several varieties of GAs (Brindle, 1981; Baker, 1985, 1987; Goldberg et al., 1991). The elitist version, which allows the best individual(s) from a generation to carry over to the next one, was introduced by De Jong (1975). Other versions are the modified GA (modGA) (Michalewicz, 1996), messy GAs (Goldberg et al., 1990), GAs with varying population size (GAsVaPS) (Michalewicz, 1996), genetic implementor (GENITOR) (Whitley, 1989), and breeder GAs (BGA) (Muhlenbein and Schlierkamp, 1993). Several authors have implemented the GA in water resources optimization (East and Hall, 1994; Gen and Cheng, 1997). Furuta et al. (1996) presented a decision-making supporting system based on the GA for the aesthetic design of dams. Pillay et al. (1997) applied genetic algorithms to the problem of parameter determination of induction motors. Wardlaw and Sharif (1999) employed the GA to solve four- and ten-reservoir problems. Several researchers implemented the GA to design flood control systems (Shafiei and Bozorg-Haddad, 2005; Shafiei et al., 2005; Bozorg-Haddad et al., 2015). Saadatpour et al. (2005) developed a simulation–optimization model based on the GA to calculate the best compromise

Meta-Heuristic and Evolutionary Algorithms for Engineering Optimization,
First Edition. Omid Bozorg-Haddad, Mohammad Solgi, and Hugo A. Loáiciga.

solution for wasteload allocation. Bozorg-Haddad et al. (2005) implemented the GA in the optimal design of stepped spillways of dams. Hosseini et al. (2010) presented an optimization model based on the GA to design rainfall gage networks. Rasoulzadeh-Gharibdousti et al. (2011) presented a hybrid GA for the optimal design and operation of pumping stations. Fallah-Mehdipour et al. (2013) applied evolutionary algorithms including the GA, particle swarm optimization (PSO), and shuffled frog leaping algorithm (SFLA) for calculating multi-crop planning rules in a reservoir system. Khorsandi et al. (2014) applied optimization techniques including the GA and the pattern search (PS) for locating and quantifying water–surface pollutants. Bhoskar et al. (2015) reported a literature review of applications of the GA in mechanical engineering. Montaseri et al. (2015) developed a simulation–optimization model based on the GA for urban stormwater management.

4.2 Mapping the Genetic Algorithm (GA) to Natural Evolution

The basic idea behind the GA is the Darwinian principle of survival of the fittest among organisms threatened by predators and environmental hazards. The fittest members have a better chance of survival than others. They are more likely to adapt to evolving conditions, and their offspring may inherit their traits and learn their skills, thus producing even fitter future generations. Furthermore, genetic mutations occur randomly in members of species, and some of those mutations may improve the chances of long-term persistence of fit individuals and their evolutionary descendants. Each individual generated by the GA (called chromosome) plays the role of a possible solution of the optimization problem at hand. Each chromosome is made up of genes that represent decision variables. The fitness values of individuals determine their ability to survive. Each generation contains a mixture of a parent population, which contains surviving individuals (chromosomes) from previous generation, and their children. The offsprings or children, which represent new solutions, are generated by genetic operators including crossover and mutation. Parents are chosen to generate a new generation so that their probability of selection is proportionate to their fitness values. The higher the fitness value, the better the chance to survive and reproduce. Table 4.1 lists the characteristics of the GA.

Standard GA begins with a randomly generated population of possible solutions (individuals). The individuals' fitness is calculated, and some of them are selected as parents according to their fitness values. A new population (or generation) of possible solutions (the children's population) is produced by applying the crossover operator to the parent population and then applying the mutation operator to their offspring. The iterations involving the replacement of the original generation (old individual) with a new generation (children) are repeated until the stopping criteria are satisfied. Figure 4.1 illustrates the flowchart of the GA.

Table 4.1 The characteristics of the GA.

General algorithm (see Section 2.13)	Genetic algorithm
Decision variable	Gene of chromosome
Solution	Chromosome (individual)
Old solution	Parent
New solution	Children (offspring)
Best solution	Elite
Fitness function	Quality of individual
Initial solution	Random chromosome
Selection	Surviving parents
Process of generating new solution	Genetic operators

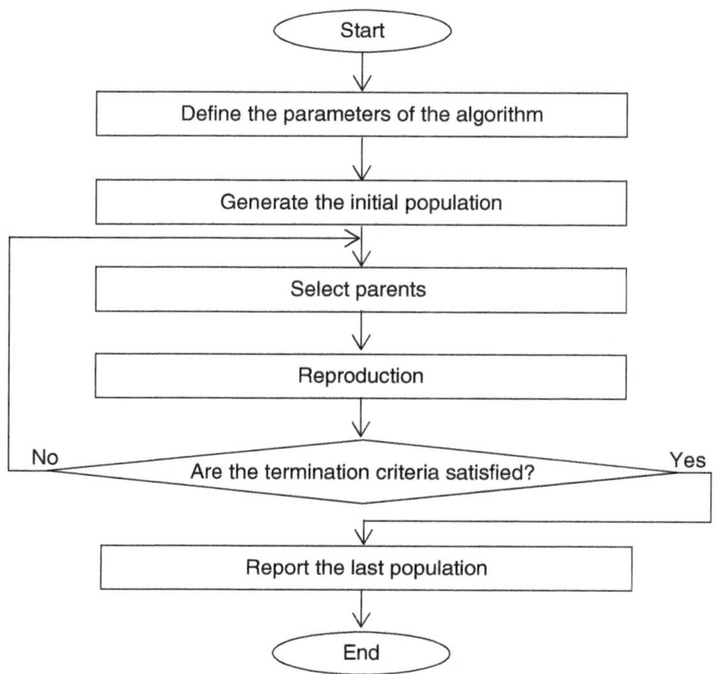

Figure 4.1 The flowchart of the GA.

4.3 Creating an Initial Population

Each possible solution of the optimization problem generated by the GA is called a chromosome. Therefore, in the mathematical formulation of an optimization problem, each chromosome is made up of a series of genes (decision variables) that represent a possible solution of the optimization problem at hand. In an N-dimensional optimization problem, a chromosome is an array of size $1 \times N$. This array is defined as follows:

$$Chromosome = X = \left(x_1, x_2, \ldots, x_i, \ldots, x_N\right) \tag{4.1}$$

where X = a possible solution of the optimization problem, x_i = ith decision variable (or gene) of solution X, and N = number of decision variables. The GA begins with random generation of a population of chromosomes or possible solutions (see Section 2.6). The population size, or the number of possible solutions, is denoted by M. The population of generated possible solutions is represented as a matrix of chromosomes of size $M \times N$:

$$Population = \begin{bmatrix} X_1 \\ X_2 \\ \vdots \\ X_j \\ \vdots \\ X_M \end{bmatrix} = \begin{bmatrix} x_{1,1} & x_{1,2} & \cdots & x_{1,i} & \cdots & x_{1,N} \\ x_{2,1} & x_{2,2} & \cdots & x_{2,i} & \cdots & x_{2,N} \\ & & & \vdots & & \\ x_{j,1} & x_{j,2} & \cdots & x_{j,i} & \cdots & x_{j,N} \\ & & & \vdots & & \\ x_{M,1} & x_{M,2} & \cdots & x_{M,i} & \cdots & x_{M,N} \end{bmatrix} \tag{4.2}$$

in which X_j = jth solution (or chromosome), $x_{j,i}$ = ith decision variable (or gene) of the jth solution, and M = population size. Each decision variable $x_{j,i}$ can be represented as a floating point number (real values) or as a predefined set of values for discrete problems. Some of the initially generated possible solutions are selected as parents to produce a new generation.

4.4 Selection of Parents to Create a New Generation

Selection in the GA is the procedure by which R ($R < M$) individuals are chosen from the population for reproduction. The selected individuals are the parents of the next generation and constitute the parent population. There are different methods for selection of the parents. The most common methods are proportionate selection, ranking selection, and tournament selection.

4.4.1 Proportionate Selection

A popular selection approach is proportionate selection (Goldberg, 1989). According to proportionate selection the probability of a solution being selected is evaluated as follows:

$$P_k = \frac{F(X_k)}{\sum_{j=1}^{M} F(X_j)} \tag{4.3}$$

in which P_k = the probability of solution kth being selected and $F(X)$ = the fitness function of solution X.

First, the probability of selection of each solution is evaluated. Each solution k belongs to the parent population with probability P_k. Based on the evaluated probabilities, a roulette wheel is made and turned (i.e., set up numerically and calculated thereupon) to select parents. The concept of a roulette wheel is depicted in Figure 4.2 with a simple example having a population of three individuals. Each individual (solution) possesses a part of the roulette wheel that is proportionate to its fitness value (F). The roulette wheel is made and spun to select a parent. Selection is biased toward fitter individuals even though it is random and any individual has a chance to be selected.

A roulette wheel is created by calculating a cumulative probability for all solutions as follows:

$$Q_j = \sum_{k=1}^{j} P_k, \quad j = 1,2,\ldots,M \tag{4.4}$$

in which Q_j = cumulative probability of the jth solution.

The selection of R parents following the creation of the roulette wheel is accomplished by spinning the wheel R times. Each spin is tantamount to a generated random number (*Rand*) in the range [0,1]. If *Rand* is less than Q_1, the first solution (X_1) is selected; otherwise the jth solution is selected such that *Rand* is greater than Q_{j-1} and less or equal than Q_j ($Q_{j-1} < Rand \le Q_j$).

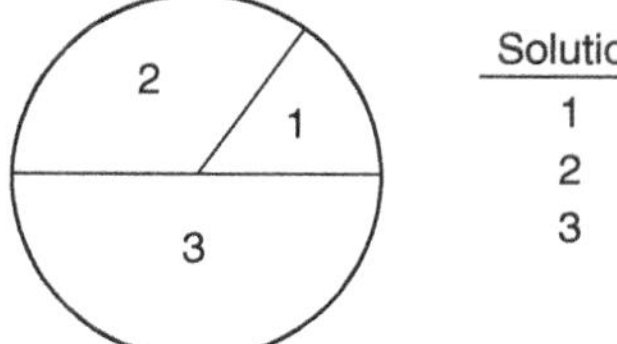

Solution	F	P
1	60	0.50
2	40	0.33
3	20	0.17

Population size (M) = 3

Figure 4.2 Demonstration of a roulette wheel. F = Fitness function value; P = probability of selecting a solution.

4.4.2 Ranking Selection

Ranking selection ranks all the solutions (also called chromosomes) based on their fitness values. The best solution receives rank 1, and the worst is assigned the lowest rank. A solution is assigned a probability that is proportionate to its rank according to the following linear function:

$$P_k = U - (S_k - 1) \times Z \tag{4.5}$$

$$S_k = Rank(X_k) \tag{4.6}$$

$$\sum_{j=1}^{M} P_j = 1 \tag{4.7}$$

$$U = \frac{Z(M-1)}{2} + \frac{1}{M} \tag{4.8}$$

in which S_k = the rank of the kth solution in the population, the term $S_k = 1$ indicates that the kth solution is the best solution, and Z = a user-defined value. Figure 4.3 depicts the sorting of solutions according to the fitness function (F) in a maximizing problem.

An alternative approach ranks all the solutions according to their fitness values. Then $M - S$ copies of each solution are generated. For example, in a population of ten solutions ($M = 10$), for a solution of rank $S = 3$, $10 - 3 = 7$ copies are made. R parent solutions are selected using the uniform distribution from the population, which is a mixture of the original solutions and their copies. The probability of choosing the fitter (better) solutions would be higher than those of less fit solutions due to their larger number of copies.

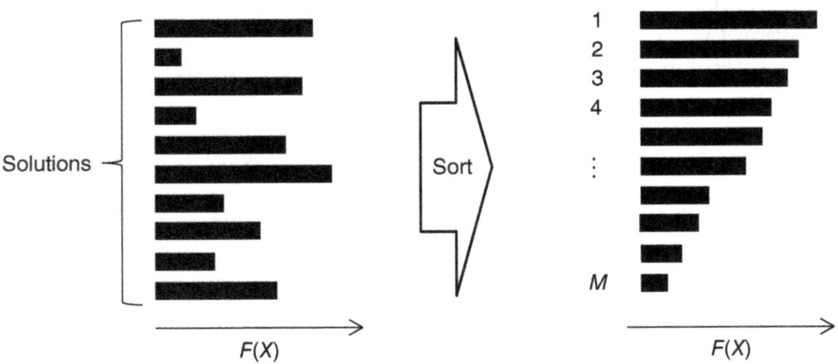

Figure 4.3 Ranking chromosomes (or solutions) according to the fitness function (F) in a maximizing problem.

4.4.3 Tournament Selection

Another popular selection method is tournament selection. According to the tournament selection Y ($Y < M$), solutions are randomly selected with the uniform distribution. When applying the uniform distribution, the probability of selecting any solution is the same. The best solution in the selected set is chosen as a parent. This process is repeated until all parents are selected.

4.5 Population Diversity and Selective Pressure

Population diversity and selective pressure are important factors in the search process of the GA. These factors are inversely related so that increasing one reduces the other (Whitley, 1989). A high selective pressure may lead to premature convergence, while a low selective pressure may lead to stagnation in the optimization search.

Proportionate selection, discussed in a previous subsection, may introduce convergence errors. Premature convergence of the GA may occur when the probability P of selecting a solution is estimated from its fitness function value if there are large differences between the fitness values of the solutions of a population, or stagnation may occur if there are small differences between the fitness function values of the solutions (Whitley, 1989; Michalewicz, 1996). Several scaling functions have been introduced to strike a balance between selective pressure and population diversity. These scaling functions are employed with proportionate selection and include linear scaling (Michalewicz, 1996), sigma truncation (Michalewicz, 1996), power law scaling (Michalewicz, 1996), logarithmic scaling (Grefenstette and Baker, 1989), exponential scaling (Grefenstette and Baker, 1989), and Boltzmann selection (Back, 1994). Solgi et al. (2016) proposed a new scaling function that is self-regulating and does not have parameters to be adjusted. This departs from previous scaling methods that require the analyst to set their parameters to regulate the selective pressure.

Ranking selection allows the users to adjust the selective pressure of the algorithm. If $Z = 0$ (and consequently, $U = 1/M$), there is no selection pressure. In this condition, all individuals have the same probability of selection. The maximum selective pressure is achieved when $U = 2/M$ and $Z = 2/(M(M-1))$. Large values of Y increase the selective pressure in tournament selection. A value $Y = 2$ is commonly used.

4.6 Reproduction

The GA algorithm must generate new solutions to progress toward an optimal solution. The parents make children that constitute the whole or a part of the next generation. Therefore, the next generation may be a mixture of the parent

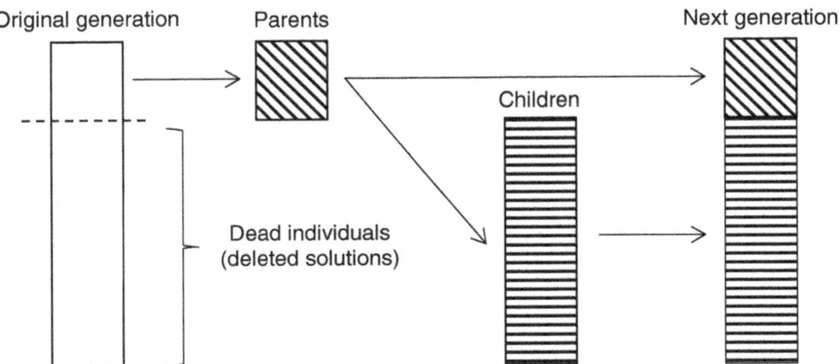

Figure 4.4 The process of constituting a new generation from the previous generation.

population and the children population. The ratio of parents to offspring is a user-defined parameter. Figure 4.4 gives an example of the procedure of producing the next generation from the original generation. Different methodologies have been devised for constituting the next generation, which define varieties of the GA.

Generating new solutions is the duty of the selected parents. Children (offspring) are new solutions. *R* solutions are selected as the population of parents based on their fitness. From this population parents are randomly selected once more with a crossover probability (P_C) that is a user-defined parameter of the algorithm. A random number *Rand* from the range [0,1] is generated for each solution in the parent population. If *Rand* is less than P_C, that solution is selected for crossover. Not all parent solutions generate children. Among the selected parent solutions, some are chosen pairwise with the uniform distribution to produce offspring. This process is called crossover. Offspring or children, which constitute solutions, are modified by the mutation operator. Thus, the GA first generates children by crossover and modifies them by mutation thereafter.

4.6.1 Crossover

Crossover occurs between two parent solutions. The crossover operator generates new offspring by exchanging genes between parents. According to the crossover operation, some decision variables of two solutions are exchanged. In other words, a new solution receives some decision variables from one parent solution and the rest from the other parent solution. Goldberg (1989) and Michalewicz (1996) have described several methods of crossover including (1) one-point crossover, (2) two-point crossover, and (3) uniform crossover. Figure 4.5 illustrates the latter three types of crossover.

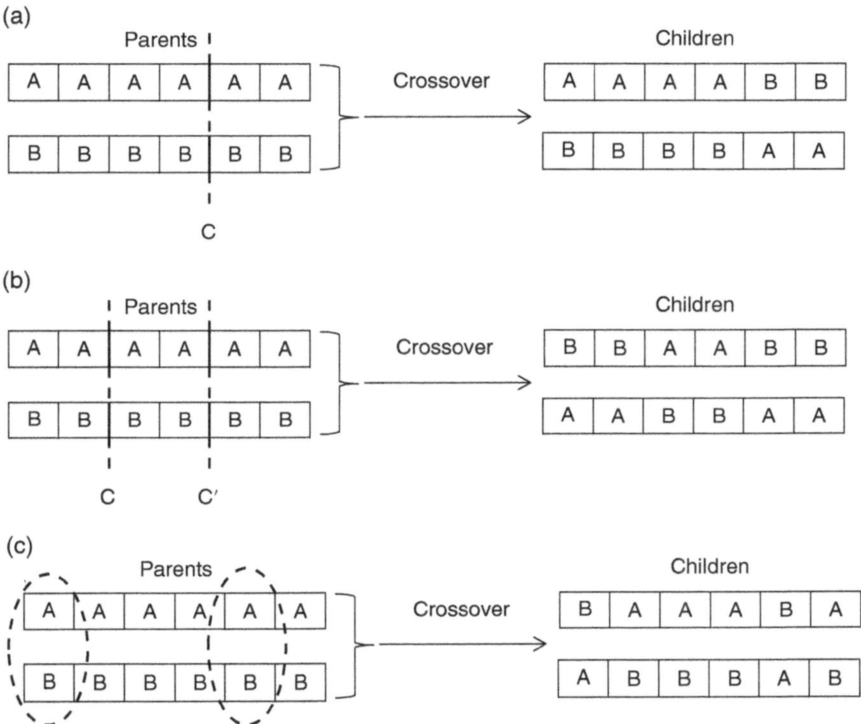

Figure 4.5 Different approaches of crossover: (a) one-point crossover, (b) two-point crossover, and (c) uniform crossover.

A crossover point (C in Figure 4.5a) is randomly selected when employing one-point crossover. A child is generated so that some of its genes are those from one of the parents and are located on one side of the point, and the rest of its gens come from the other parent and are located on the other side of the point as shown in Figure 4.5a. Each couple of parents generates two children. Two crossover points are randomly generated when employing two-point crossover and are denoted by C and C′ in Figure 4.5b. The genes between the two points in the parent solutions are preserved in the same positions in the genetic make of the children. The genes positioned outside the two boundaries are exchanged as shown in Figure 4.5b to produce the two children. The scheme for uniform crossover is depicted in Figure 4.5c and is self-explanatory. Crossover points are generated as an integer random number in the range $[1,N]$. To illustrate one-point crossover generation of two children from two

N-dimensional parent solutions $X=(x_1,x_2,\ldots,x_N)$ and $X'=(x'_1,x'_2,\ldots,x'_N)$, let C denote the crossover point. Therefore, the children are generated as follows:

$$X_1^{(new)}=(x_1,x_2,\ldots,x_c,x'_{c+1},x'_{c+2},\ldots,x'_N) \tag{4.9}$$

$$X_2^{(new)}=(x'_1,x'_2,\ldots,x'_c,x_{c+1},x_{c+2},\ldots,x_N) \tag{4.10}$$

in which $X^{(new)}$ = new solution.

4.6.2 Mutation

Mutation is important because it introduces new genetic material to a population. The mutation operator replaces randomly some genes of an offspring. In other words, one or more decision variables of a new solution are replaced with random values while keeping the values of its other decision variables unaltered. Figure 4.6 illustrates the mutation operator.

Two methods of mutation for real-valued variables are uniform and nonuniform mutations. Uniform mutation replaces a parent gene with a randomly generated gene that is within the feasible space of the solutions. Let $X=(x_1,x_2,\ldots,x_i,\ldots,x_N)$ and x_i denote a solution (chromosome) and a gene (decision variable), respectively, where the decision variable ith (x_i) is chosen for mutation. Uniform mutation produces a mutated $X'=(x_1,x_2,\ldots,x'_i,\ldots,x_N)$ whereby x'_i is evaluated as follows:

$$x'_i=Rnd\left(x_i^{(L)},x_i^{(U)}\right) \tag{4.11}$$

in which x'_i = the new value of x_i produced by mutation, $x_i^{(U)}$ = the upper bound of the ith decision variable, $x_i^{(L)}$ = the lower bound of the ith decision variable, and $Rnd(a,b)$ = a random value chosen from the feasible range $[a,b]$.

Nonuniform mutation induces an increasingly localized search for optimal solutions in which the sets of genes that are chosen for mutation are defined by means of boundaries that become narrower as the run of the GA progresses (Michalewicz, 1996). This type of mutation is especially useful for problems in

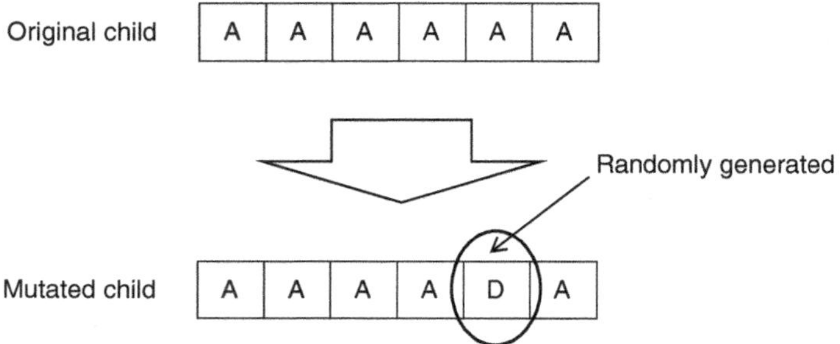

Figure 4.6 An example of the mutation operator.

which high precision is required. Let $X = (x_1, x_2, \ldots, x_i, \ldots, x_N)$ be a solution of an optimization problem and its ith decision variable (x_i) be selected for mutation. Nonuniform mutation produces a mutated solution $X' = (x_1, x_2, \ldots, x'_i, \ldots, x_N)$ whereby x'_i is calculated as follows:

$$x'_i = Rnd(x_i - d, x_i + d) \tag{4.12}$$

$$d = d_0 \times \frac{T - t}{T} \tag{4.13}$$

in which d_0 = initial value of d, t = current iteration, and T = maximum number of iterations.

Mutation is performed probabilistically. A mutation probability (P_M) is specified that permits random mutations to be made to individual genes. The implementation of the mutation operator is applied to each decision variable of the solutions in the children population by generating a random number *Rand* in the range [0,1]. If *Rand* is less than P_M, that decision variable is mutated; otherwise it remains unaltered.

4.7 Termination Criteria

The termination criteria determine when to end the algorithm's iterations. Selecting a good termination criterion has an important role on the correct convergence of the algorithm. The number of iterations, the amount of improvement of the objective function between consecutive iterations, and the run time are common termination criteria for the GA.

4.8 User-Defined Parameters of the GA

The size of the population of solutions (M), the number of parents (R), the probability of crossover (P_C), the probability of mutation (P_M), and the termination criterion are the user-defined parameters of the GA. A good choice of the parameters is related to the decision space of a particular problem, and in general the optimal parameter setting for one problem may not perform equally as well for other problems. Consequently, determining a good parameter setting often requires the execution of a large number of time-consuming experiments. Mastering the choice of the GA parameters relies on practice and experience with specific optimization problems. However, a reasonable method for finding suitable values for the parameters is performing sensitivity analysis. This entails choosing a combination of parameters and running the GA several times. Other combinations of parameters are chosen, and repeated runs are made with each combination. A comparison of the optimization results so obtained sheds light on the best set of GA parameters.

4.9 Pseudocode of the GA

```
Begin
  Input parameters of the algorithm and initial data
  Let R and M = number of parents and the size of
    population, respectively
  Generate M initial possible solutions
  While (the termination criteria are not satisfied)
     Evaluate fitness values for all solutions
     Select the parent population with a selection
       method
     For j = 1 to R
         Generate a random value Rand in the range [0,1]
         If Rand < P_C
             Parent j is known as an effective solution
         Otherwise
             Parent j is known as an ineffective solution
         End if
     Next j
     For j = 1 to (M - R) / 2
         Select two solutions randomly with the uniform
           distribution from effective parents of the
           parent population.
         Generate two new solutions with the crossover
           operator
         Put new generated solutions into the children
           population
     Next j
     For j = 1 to (M - R)
         For i = 1 to N
             Generate a random value Rand in the
               range [0,1]
             If Rand < P_M
                 Replace the decision variable i from
                   solution j (x_j,i) using the mutation
                   operator
             End if
         Next i
     Next j
     Set population = parent population + children
       population
  End while
  Report the population
End
```

4.10 Conclusion

This chapter described the GA, which is a well-known evolutionary algorithm. First, a brief literature review of the GA was presented. The natural evolutionary process was mapped to the GA, and its key components were described. A pseudocode of the GA closed the GA review.

References

Back, T. (1994). "Selective pressure in evolutionary algorithms: A characterization of selection mechanisms." IEEE World Congress on Computational Intelligence, Orlando, FL, Jun 27–29, Piscataway, NJ: Institute of Electrical and Electronics Engineers (IEEE), 57–62.

Baker, J. E. (1985). "Adaptive selection methods for genetic algorithms." In: Grefenstette, J. J. (Ed.), Proceedings of the first international conference on genetic algorithms, July 28–31, Lawrence Erlbaum Associates, Hillsdale, NJ, 101–111.

Baker, J. E. (1987). "Reducing bias and inefficiency in the selection algorithm." In: Grefenstette, J. J. (Ed.), Proceedings of the second international conference on genetic algorithms, July 28–31, Lawrence Erlbaum Associates, Hillsdale, NJ, 14–21.

Bhoskar, T., Kulkarni, A. K., Kulkarni, N. K., Patekar, S. L., Kakandikar, G. M., and Nandedkar, V. M. (2015). "Genetic algorithm and its applications to mechanical engineering: A review." Materials Today: Proceedings, 2(4), 2624–2630.

Bozorg-Haddad, O., Ashofteh, P., and Mariño, M. A. (2015). "Levee layouts and design optimization in protection of flood areas." Journal of Irrigation and Drainage Engineering, 141(8), 04015004.

Bozorg-Haddad, O., Sharifi, F., and Naderi, M. (2005). "Optimum design of stepped spillways using genetic algorithm." Proceedings of the 6th WSEAS International Conference on Evolutionary Computing, Lisbon, Portugal, June 16–18, Madison, WI: University of Wisconsin, 325–331.

Brindle, A. (1981). "Genetic algorithms for function optimization." Doctoral Dissertation, University of Alberta, Edmonton.

De Jong, K. A. (1975). "An analysis of the behavior of a class of genetic adaptive systems." Doctoral Dissertation, University of Michigan, Ann Arbor, MI.

East, V. and Hall, M. J. (1994). "Water resources system optimization using genetic algorithms." In: Verwey, A., Minns, A. W., Babovic, V., and Maksimovic, C. (Eds.), Proceeding of the 1st international conference on hydroinformatics, Delft, Netherlands, September 19–23, A.A. Balkema, Rotterdam.

Fallah-Mehdipour, E., Bozorg-Haddad, O., and Mariño, M. A. (2013). "Extraction of multicrop planning rules in a reservoir system: Application of evolutionary algorithms." Journal of Irrigation and Drainage Engineering, 139(6), 490–498.

Furuta, H., Hase, H., Watanabe, E., Tonegawa, T., and Morimoto, H. (1996). "Applications of genetic algorithm to aesthetic design of dam structures." Advances in Engineering Software, 25(2–3), 185–195.

Gen, M. and Cheng, R. W. (1997). "Genetic algorithms and engineering design." John Wiley & Sons, Inc., New York.

Goldberg, D. E. (1989). "Genetic algorithms in search, optimization and machine learning." Addison-Wesely, Boston, MA.

Goldberg, D. E., Deb, K., and Korb, B. (1991). "Do not worry, be messy." In: Belew, R. K. and Booker, L. B. (Eds.), Proceedings of the fourth international conference on genetic algorithms, July 24–30, Morgan Kaufman, San Mateo, CA.

Goldberg, D. E., Deb, K., and Krob, B. (1990). "Messy genetic algorithms revisited: Nonuniform size and scale." Complex Systems, 4(4), 415–444.

Grefenstette, J. J. and Baker, J. E. (1989). "How genetic algorithms work: A critical look at implicit parallelism." In: Schaffer, J. D. (Ed.), Proceedings of the third international conference on genetic algorithms, July 20–27, Morgan Kaufman, San Mateo, CA.

Holland, J. H. (1975). "Adaptation in natural and artificial systems." University of Michigan Press, Ann Arbor, MI.

Hosseini, A. K., Bozorg-Haddad, O., and Shadkam, S. (2010). "Rainfall network optimization using transinformation entropy and genetic algorithm." 21st Century Watershed Technology: Improving Water Quality and Environment Conference Proceedings, Universidad EARTH, Mercedes, CR, February 21–24, St. Joseph, MI: American Society of Agricultural and Biological Engineers (ASABE).

Khorsandi, M., Bozorg-Haddad, O., and Mariño, M. A. (2014). "Application of data-driven and optimization methods in identification of location and quantity of pollutants." Journal of Hazardous, Toxic, and Radioactive Waste, 19(2), 04014031.

Michalewicz, Z. (1996). "Genetic algorithms + data structures = evolution programs." Springer-Verlag, Berlin, Heidelberg.

Montaseri, M., Afshar, M. H., and Bozorg-Haddad, O. (2015). "Development of simulation-optimization model (MUSIC-GA) for urban stormwater management." Water Resources Management, 29(13), 4649–4665.

Muhlenbein, H. and Schlierkamp, P. (1993). "Predictive models for the breeder genetic algorithm." Evolutionary Computation, 1(1), 25–49.

Pillay, P., Nolan, R., and Haque, T. (1997). "Application of genetic algorithms to motor parameter determination for transient torque calculations." Transactions on Industry Applications, 33(4–5), 1273–1282.

Rasoulzadeh-Gharibdousti, S., Bozorg-Haddad, O., and Mariño, M. A. (2011). "Optimal design and operation of pumping stations using NLP-GA." Proceedings of the Institution of Civil Engineers-Water Management, 164(4), 163–171.

Saadatpour, M., Afshar, A., and Bozorg-Haddad, O. (2005). "Simulation-optimization model for fuzzy waste load allocation." Proceedings of the 6th WSEAS International Conference on Evolutionary Computing, Lisbon, Portugal, June 16–18, Madison, WI: University of Wisconsin, 384–391.

Shafiei, M. and Bozorg-Haddad, O. (2005). "Optimization of levee's setback: A new GA approach." Proceedings of the 6th WSEAS International Conference on Evolutionary Computing, Lisbon, Portugal, June 16–18, Madison, WI: University of Wisconsin, 400–406.

Shafiei, M., Bozorg-Haddad, O., and Afshar, A. (2005). "GA in optimizing Ajichai flood Levee's encroachment." Proceedings of the 6th WSEAS International Conference on Evolutionary Computing, Lisbon, Portugal, June 16–18, Madison, WI: University of Wisconsin, 392–399.

Solgi, M., Bozorg-Haddad, O., and Loáiciga, H. A. (2016). "The enhanced honey-bee mating optimization algorithm for water resources optimization." Water Resources Management, 31, 885.

Wardlaw, R. and Sharif, M. (1999). "Evolution of genetic algorithms for optimal reservoir system operation." Journal of Water Resources Planning and Management, 125(1), 25–33.

Whitley, D. (1989). "The GENITOR algorithm and selection pressure: Why rank-based allocation of reproduction trials is best." In: Scahffer, J. D. (Ed.), Proceedings of the third international conference on genetic algorithms, June 4–7, Morgan Kaufman, San Mateo, CA, 116–121.

5

Simulated Annealing

Summary

This chapter reviews the simulated annealing (SA) algorithm. The SA is inspired by the process of annealing in metallurgy. It is one of the meta-heuristic optimization algorithms. This chapter presents a literature review of the development and applications of the SA, followed by a description of the process of physical annealing and its mapping to the SA, which outlines the steps of the algorithm in detail. The chapter closes with a pseudocode of the SA algorithm.

5.1 Introduction

A popular algorithm in heuristic optimization, simulated annealing (SA) optimization was developed by Kirkpatrick et al. (1983), who showed how a model for simulating the annealing of solids, as proposed by Metropolis et al. (1953), could be used for solving optimization problems in which the fitness or objective function to be minimized corresponds to the energy states of the solid. Dolan et al. (1989) demonstrated the capacity of the SA for optimizing chemical processes by applying it to the design of pressure relief header networks and heat exchanger networks. Dougherty and Marryott (1991) described the SA algorithm and applied it to the optimization of groundwater management problems in combinatorial form. Wang and Zheng (1998) linked the SA with MODFLOW, a groundwater flow simulation code, for optimal management of groundwater resources. The results of the SA were compared with those obtained with linear programming, nonlinear programming, and differential dynamic programming. The comparison showed better solutions by the SA than by other methods. Cunha and Sousa (1999) implemented the SA to obtain the least-cost design of a looped water distribution network

Meta-Heuristic and Evolutionary Algorithms for Engineering Optimization,
First Edition. Omid Bozorg-Haddad, Mohammad Solgi, and Hugo A. Loáiciga.

and proved the ability of the SA to handle this kind of problems. Cunha (1999) applied the SA for solving aquifer management problems. Ceranic et al. (2001) implemented the SA algorithm to the minimum-cost design of reinforced concrete retaining structures. Tospornsampan et al. (2005) demonstrated good performance of the SA in solving a ten-reservoir optimization problem. Alkandari et al. (2008) applied the SA algorithm to electric power quality analysis. Orouji et al. (2013) compared the performance of the SA with that of the shuffled frog leaping algorithm (SFLA) in estimating the Muskingum flood routing parameters. Yeh et al. (2013) applied the SA and the tabu search (TS) to the optimization of sewer network designs, which are complex nonlinear problems, and reported that the performance of the SA was better than those of other methods previously reported in the literature.

5.2 Mapping the Simulated Annealing (SA) Algorithm to the Physical Annealing Process

The SA algorithm emulates the physical annealing of solids to solve optimization problems. SA is so named because of its similarity to the process of physical annealing of solids, in which a solid is heated and then cooled slowly until it attains its most possible regular crystal lattice arrangement free of crystal faults.

Annealing in metallurgy and materials science defines a process in which heat changes the physical and sometimes chemical features of a substance to increase its ductility and reduce its hardness. The particles of a solid have geometric configuration that corresponds to the minimum energy arrangement in its most stable state, as it is experimentally seen in the crystals of a mineral. Physical annealing is the process whereby the low energy arrangement of a solid is achieved by melting a substance followed by lowering its temperature slowly. The annealing of a substance involves heating it to its recrystallization temperature, maintaining a suitable temperature, followed by slowly cooling of the substance until reaching its freezing point. Annealing is a well-known process to grow crystals from a molten substance. Through the process of annealing, atoms move in the crystal lattice and the number of dislocations decreases. The annealing process changes the ductility and hardness of substances. Especially important is that in the cooling stage, the temperature of the substance has to be decreased slowly. Otherwise the resulting solid is frozen into a metastable glass or a crystal with faults in its structure.

The Metropolis Monte Carlo method was proposed to simulate the annealing process. Metropolis et al. (1953) presented a simple algorithm to simulate a system composed of a set of atoms of a substance at a specific temperature.

At each iteration an atom undergoes a small random movement, which changes the energy of the system. The resulting change and movement is accepted if the energy of the system decreases. On the other hand, the movement of the atom is accepted with probability $\exp(-\delta/k_B\lambda)$ if the movement of the atom increases the energy of the system, where λ is the temperature, k_B denotes the Boltzmann constant, and δ is the change in the energy of the system. The system achieves thermal equilibrium at each temperature after a large number of atomic movements take place at each temperature. At thermal equilibrium the probability distribution of the system states follow a Boltzmann distribution whereby the probability of the system being in state i at temperature λ equals $\exp(-E_i/k_B\lambda)/Z(\lambda)$, where E_i is the energy of state i and $Z(\lambda)$ denotes the partition function that is required for normalization. The temperature of the substance is decreased slowly to allow the system attain thermal equilibrium. The procedure presented by Metropolis et al. (1953) guarantees that the system evolves into the required Boltzmann distribution (Eglese, 1990).

Different states of a substance represent different solutions of the optimization problem when applying SA. The energy of the substance is equivalent to the fitness function to be optimized. The movement of atoms introduces a new solution. In other words, a new state of the substance is a new solution. Atom movements that introduce better solutions are accepted in SA. Non-improving (inferior) changes that result in a worse solution are accepted probabilistically. The probability of accepting non-improving solutions depends on an acceptance function. Table 5.1 lists the characteristics of the SA.

Table 5.1 The characteristics of the SA.

General algorithm (see Section 2.13)	Simulated annealing
Decision variable	Position of the atoms of a substance
Solution	State of the substance
Old solution	Current state of the substance
New solution	New state of the substance
Best solution	–
Fitness function	Energy of the substance
Initial solution	Random position
Selection	Acceptance function
Process of generating new solution	Movement of atoms

The SA starts by generating an initial random solution that is known as the current state of the substance. A new state in the neighborhood of the current state is generated by a suitable mechanism, and the fitness value of the new state is evaluated. The new state is accepted if it is better than the current state. Otherwise, the new state is inferior to the current state, and the new state is accepted if it is successful according to a probabilistic function called the acceptance function, which depends on the system's temperature. The algorithm proceeds by generating a certain number of new states at each temperature while the temperature is gradually decreased. Several new solutions are tried until a thermal equilibrium criterion is satisfied at each temperature. At that juncture the temperature is decreased again. As the temperature is reduced, the probability of selecting non-improving atom movements is also reduced. The process of generating new solutions as the system is cooled is repeated until termination criteria are satisfied. Figure 5.1 illustrates the flowchart of the SA algorithm.

5.3 Generating an Initial State

Each possible solution of an optimization problem generated by the SA corresponds to an arrangement of the atoms of a substance. The state of the system consists of N decision variables in an N-dimensional space. The system state is represented as an array of size $1 \times N$ describing the arrangement of the atoms of the substance. The SA starts with a single solution denoted by an array of size $1 \times N$ as follows:

$$State = X = \left(x_1, x_2, \ldots, x_i, \ldots, x_N\right) \tag{5.1}$$

where X = a solution of the optimization problem, x_i = ith decision variable of solution X, and N = the number of decision variables. The decision variable values ($x_1, x_2, x_3, \ldots, x_N$) can be represented as a floating point number (real values) or as a predefined sets of values for continuous and discrete problems, respectively. An initial state is randomly generated to start the optimization algorithm (see Section 2.6). That initial state becomes the current state of the system (i.e., a substance or a solid).

5.4 Generating a New State

Atoms move to new places to decrease the energy of a system and achieve a sustainable state during annealing. A new potential state of the system is generated according to the current state. Several deterministic and stochastic schemes are

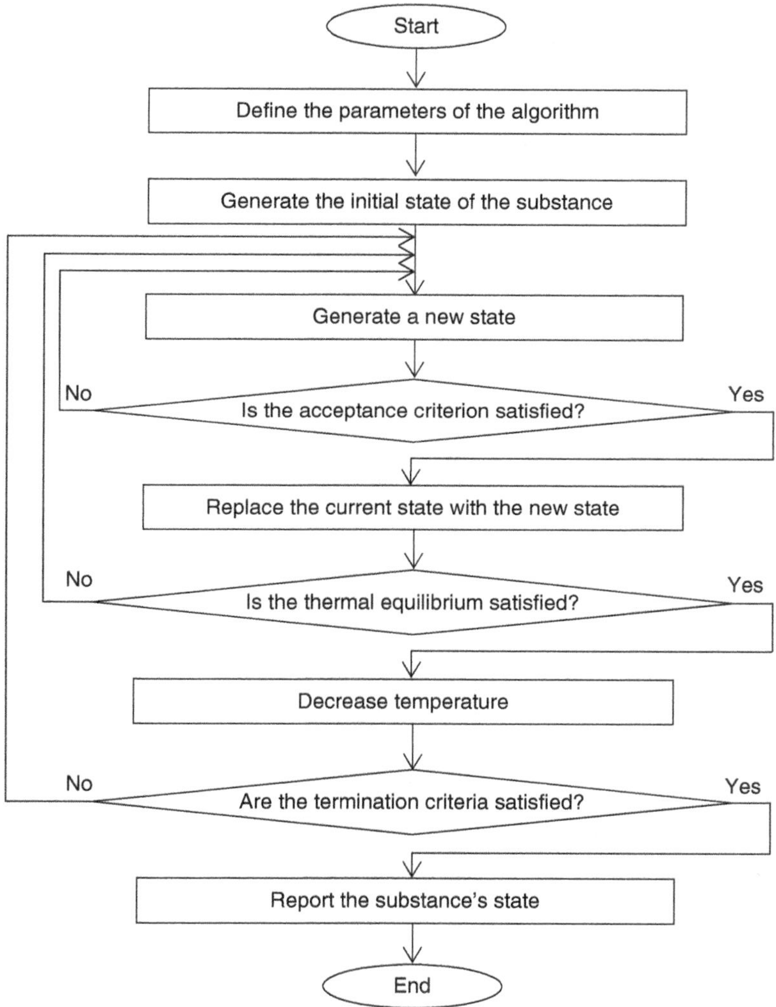

Figure 5.1 The flowchart of the SA.

used to generate a neighbor solution from a given solution. One common scheme is the random walk, which is expressed mathematically as follows:

$$x_i' = x_i + Rnd(-\varepsilon,\varepsilon), \quad i = 1,2,\ldots,N \tag{5.2}$$

in which x_i' = new value of decision variable ith, $Rnd(a,b)$ = a random value from range [a,b], and ε = a small value.

New values of all the decision variables are evaluated prior to generating a new solution. Then, a new solution is generated as follows:

$$X^{(new)} = \left(x_1', x_2', \ldots, x_i', \ldots, x_N'\right) \tag{5.3}$$

in which $X^{(new)}$ = new solution.

A newly generated solution may replace an old solution based on the acceptance function, which is described next.

5.5 Acceptance Function

The acceptance function determines whether or not a newly generated solution replaces the current solution. The new solution represents a potential arrangement of the substance, and its acceptance or rejection is based on the fitness values of the old solution and newly generated solution. The new solution replaces the old solution whenever its fitness value is superior to that of the old solution. Otherwise, the new solution is replaced by the old solution with a specific probability that is calculated based on the difference between the fitness values of the new and the old solutions. In the case of a minimization problem, the new solution is accepted according to an acceptance probabilistic function as follows:

$$P\left(X, X^{(new)}\right) = \begin{cases} 1 & if \quad F\left(X^{(new)}\right) < F(X) \\ e^{\frac{-\Delta F}{\lambda}} & Otherwise \end{cases} \tag{5.4}$$

$$\Delta F = \left| F\left(X^{(new)}\right) - F(X) \right| \tag{5.5}$$

in which X = the old solution, $X^{(new)}$ = the newly generated solution, $P(X, X^{(new)})$ = the probability of replacing X with $X^{(new)}$, $F(X)$ = the fitness value of solution X, and λ = a control parameter that corresponds to the temperature in the analogy with physical annealing. A uniformly distributed random variable (*Rand*) within [0,1] is generated, and $P(X, X^{(new)})$ is evaluated. If $P(X, X^{(new)})$ is larger than or equal to *Rand*, the newly generated solution ($X^{(new)}$) replaces the old solution (X); otherwise it is rejected. The acceptance function defined by Equations (5.4) and (5.5) is the Boltzmann distribution. The previously defined acceptance function implies that small differences in the fitness function value are more likely to be accepted than large differences. When λ is large non-improving changes are accepted more easily than when λ is relatively small. In other words, the selective pressure is high (low) when λ is low (high). The value of λ has an important role in the correct convergence of the algorithm. The algorithm starts with a relatively large value of λ to avoid being prematurely trapped in a local optimum, and it is gradually decreased as the algorithm progresses.

5.6 Thermal Equilibrium

A system achieves thermal equilibrium at each temperature whenever a large number of neighborhood movements occur at each temperature. It can be proved that at thermal equilibrium, the probability distribution of a system states follows a Boltzmann distribution. The SA proceeds by attempting a number of neighborhood moves at each temperature. In other words, for each value of the temperature λ, a certain number of new states are generated and accepted or discarded before λ is decreased. The new states are tested by the acceptance function. The number of new states is a parameter of SA algorithm selected by the user and is herein denoted by β. Thermal equilibrium is satisfied whenever a predefined number of new solutions (β) is generated and tested by the acceptance function.

5.7 Temperature Reduction

The system's temperature is decreased after testing a number of new states. The parameter λ controls the selective pressure, which is high (low) when λ is low (high), and plays an important role in the correct convergence of the SA algorithm. The algorithm starts with a relatively high value of λ to avoid being prematurely trapped in a local optimum, and it is gradually decreased as the algorithm progresses, as follows:

$$\lim_{t \to +\infty} \lambda_t = 0, \quad t > 0 \tag{5.6}$$

in which t = the iteration counter of the algorithm.

Two common procedures for decreasing λ are linear and geometric. The linear function modifies λ in each iteration with the following equation:

$$\lambda_t = \lambda_0 - \alpha \times t \tag{5.7}$$

$$\alpha = \frac{\lambda_0 - \lambda_T}{T} \tag{5.8}$$

in which λ_0 = initial temperature, λ_t = the (modified) temperature in iteration t, T = the total number of iterations, and α = the cooling factor.

The geometric procedure for cooling the system is as follows:

$$\lambda_t = \lambda_0 \times \alpha^t, \quad 0 < \alpha < 1 \tag{5.9}$$

in which α = the cooling factor. The advantage of the geometric function is that it does not require the specification of the maximum number of iterations of the algorithm. The stopping criterion of the algorithm could be the maximum number of iterations T, but other termination criteria can be imposed, such as run time. In this case, the value of T is unknown, and the SA algorithm continues until the stopping criteria (e.g., run time) are satisfied.

5.8 Termination Criteria

The termination criterion determines when to terminate the SA algorithm. Selecting a suitable termination criterion has an important role on the correct convergence of the algorithm. The number of iterations, the incremental improvement of the objective function between consecutive iterations, and the run time are common termination criteria applied in the implementation of the SA algorithm.

5.9 User-Defined Parameters of the SA

The initial value of λ, the value of β, which determines the number of new generated solutions at each value of λ (thermal equilibrium), the rate of decrease of λ, and the termination criterion are user-defined parameters of the SA. These parameters are known as the annealing or cooling schedule. The choice of the annealing schedule influences the performance of the algorithm. Annealing schedules that have been recommended for successful convergence have not proven successful in all practical applications. Thus, the application of the SA algorithm requires the implementation of heuristic criteria that strike an acceptable trade-off between time invested in selecting the SA parameters and the quality of the solution achieved. A good choice of the parameters depends on the decision space of the optimization problem. Frequently the optimal parameter setting for one problem is of limited utility for other problems. Consequently, determining a good parameter set often requires a large number of time-consuming experiments. The proper choice of the SA parameters involves practice and experience with the type of problem being solved. Sensitivity analysis is a reasonable method for finding appropriate values for the SA parameters. Sensitivity analysis prescribes a combination of parameters with which the SA algorithm is run for several times. Several other combinations are chosen and the algorithm is run several times with each of them. A comparison of the results calculated from many runs provides guidance about a suitable choice of the SA parameters.

5.10 Pseudocode of the SA

```
Begin
  Input parameters of the algorithm and initial data
  Generate initial solution possible X and evaluate
    its fitness value
  Let β = the number of new solutions which are generated
    to reach thermal equilibrium
  While (termination criteria are not satisfied)
      For j = 1 to β
          Generate a new solution X(new) and evaluate its
            fitness value
          If the new generated solution (X(new)) is better
            than the old one (X)
              Put X = X(new)
          Otherwise
              Evaluate P(X, X(new)) and generate Rand
                from the range [0,1] randomly
              If P(X, X(new)) > Rand
                   Put X = X(new)
              End if
          End if
      Next j
      Decrease the temperature
  End while
  Report the solution X
End
```

5.11 Conclusion

This chapter explained the SA algorithm that is inspired by the process of annealing in metal work. The physical annealing process was mapped into the SA after a brief literature review, and the steps of the algorithm were described. A pseudocode of the SA algorithm closed this review.

References

Alkandari, A. M., Soliman, S. A., and Mantwy, A. H. (2008). "Simulated annealing optimization algorithm for electric power systems quality analysis: Harmonics and voltage flickers." 12th International Middle-East (MEPCON), Aswan, Egypt, March 12–15, Piscataway, NJ: Institute of Electrical and Electronics Engineers (IEEE).

Ceranic, B., Fryer, C., and Baines, R. W. (2001). "An application of simulated annealing to the optimum design of reinforced concrete retaining structures." Computers and Structures, 79(17), 1569–1581.
Cunha, M. (1999). "On solving aquifer management problems with simulated annealing algorithms." Water Resources Management, 13(3), 153–170.
Cunha, M. and Sousa, J. (1999). "Water distribution network design optimization: Simulated annealing approach." Journal of Water Resources Management, 125(4), 215–221.
Dolan, W. B., Cummings, P. T., and LeVan, M. N. (1989). "Process optimization via simulated annealing: Application to network design." AIChE Journal, 35(5), 725–736.
Dougherty, D. E. and Marryott, R. A. (1991). "Optimal groundwater management: 1. Simulated annealing." Water Resources Research, 27(10), 2493–2508.
Eglese, R. W. (1990). "Simulated annealing: A tool for operational research." European Journal of Operational Research, 56, 271–281.
Kirkpatrick, S., Gelatt, C. D., and Vecchi, M. P. (1983). "Optimization by simulated annealing." Science, 220(4598), 671–680.
Metropolis, N., Rosenbluth, A., Rosenbluth, M., Teller, A., and Teller, E. (1953). "Equation of state calculations by fast computing machines." Journal of Chemical Physics, 21(6), 1087–1092.
Orouji, H., Bozorg-Haddad, O., Fallah-Mehdipour, E., and Mariño, M. A. (2013). "Estimation of Muskingum parameterby meta-heuristic algorithms." Proceedings of the Institution of Civil Engineers: Water Management, 165(1), 1–10.
Tospornsampan, J., Kita, I., Ishii, M., and Kitamura, Y. (2005). "Optimization of a multiple reservoir system using a simulated annealing: A case study in the Mae Klong system, Thailand." Paddy and Water Environment, 3(3), 137–147.
Wang, M. and Zheng, C. (1998). "Ground water management optimization using genetic algorithms and simulated annealing: Formulation and comparison." Journal of the American Water Resources Association, 34(3), 519–530.
Yeh, S. F., Chang, Y. J., and Lin, M. D. (2013). "Optimal design of sewer network by tabu search and simulated annealing." 2013 IEEE International Conference on Industrial Engineering and Engineering Management, Bangkok, Thailand, December 10–13, Piscataway, NJ: Institute of Electrical and Electronics Engineers (IEEE).

6

Tabu Search

Summary

This chapter describes the tabu search (TS), which is a meta-heuristic algorithm for combinatorial optimization. A brief literature review of development and applications of the TS opens this chapter. The foundations of the TS and its algorithmic steps are described next, followed by a pseudocode of the TS.

6.1 Introduction

The tabu search (TS) was developed by Glover (1986). It is designed to solve combinatorial (finite solution set) optimization problems. Simple TS and advanced TS were introduced by Glover (1989, 1990). Bland (1995) implemented the TS in a structural design context and showed that the TS is a technically viable technique for optimal structural design. Fanni et al. (1999) applied the TS coupled with deterministic strategies for the optimal design of magnetic resonance imaging (MRI) devices. Wang et al. (1999) demonstrated the capability of the TS to optimal design of multiproduct batch chemical processes. Nara et al. (2001) applied the TS to determine locations and discrete capacities of distributed generators so that the distribution loss is minimized. Hajji et al. (2004) developed a new TS algorithm for global optimization of multimodal functions with continuous domain. Misevicius (2005) implemented the TS to solve the quadratic assignment problem (QAP). Nourzad and Afshar (2009) proposed a probabilistic improvement to the neighborhood selection of the TS and used that to find optimal water resources allocation for an industrial copper complex distribution system. Hajji et al. (2010) designed software based on scatter search, the TS, and neural networks for determining water-pumping schedules. Martinez et al. (2010) employed the TS algorithm to optimize water level monitoring stations in lakes and streams within the south Florida water

Meta-Heuristic and Evolutionary Algorithms for Engineering Optimization,
First Edition. Omid Bozorg-Haddad, Mohammad Solgi, and Hugo A. Loáiciga.

management district. Yeh et al. (2013) used the TS and the SA to the optimization of sewer-network designs that are complex nonlinear problems. Haghighi and Bakhshipour (2014) presented an integrated optimization model for designing sewage collection networks and applied the TS method as a deterministic combinatorial meta-heuristic to solve the optimization model.

6.2 Tabu Search (TS) Foundation

The TS is an enhancement of local search (LS) methods. The LS refers to iterative procedures that start with a solution and then employs local modifications (moves) to find a superior solution. The basic concept of the LS is that the movement is always from a worse solution to a better one. The search terminates when it reaches an optimum with respect to the movements made. The optimum achieved with the LS is mostly a local optimum instead of a global optimum given that the algorithm always moves to an improved neighboring solution near the current one. The TS solved the problem of convergence to local optima experienced with LS methods by allowing movements to non-improving solutions when there is no better solution near the current solution. The TS also takes advantage of principles of artificial intelligence by making search movements based on memory structures that prevent repetitive movements and help to explore the decision space of the optimization problem more thoroughly. Previously visited solutions are known as tabu and moving back to them is prevented by the memories that save the history of the search for optima.

TS designates solutions of an optimization problem as points in an N-dimensional space where N denotes the number of decision variables. Neighboring points refer to new solutions. The process of going from the searching point, which is the old solution, to a neighboring point is called a move. The best point reached in the search is the best solution found during the search. Table 6.1 lists the characteristics of the TS.

A simple TS starts by generating a random solution (initial solution), which is known as the searching point. The current searching point is considered momentarily as the best point. In the next step neighboring points are generated near the searching point. Only the neighboring points that are not tabu are considered. The searching point moves to the best neighboring point that is not tabu. Unlike the LS, in the TS the best neighboring point replaces the searching point even if it is worse than the current searching point. The previous searching point is memorized as tabu. If the new searching point is better than the best point, the new point replaces the best point; otherwise the best point remains unchanged. Neighboring points near the new searching point are generated. The previously mentioned process repeats until the termination criteria are satisfied. Figure 6.1 illustrates the flowchart of the TS.

Table 6.1 The characteristics of the TS.

General algorithm (see Section 2.13)	Tabu search
Decision variable	Point's position
Solution	Point
Old solution	Searching point
New solution	Neighbor point
Best solution	Best point
Fitness function	Desirability of the point
Initial solution	Random point
Selection	Tabu list
Process of generating new solution	Movement

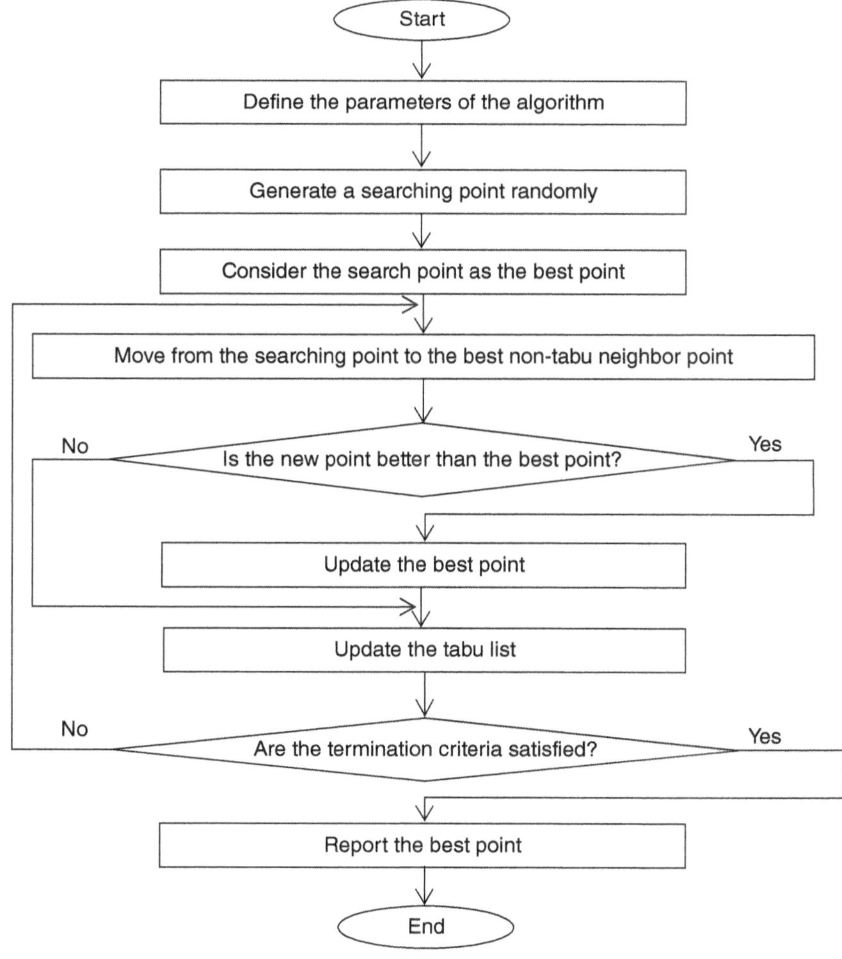

Figure 6.1 The flowchart of the TS.

6.3 Generating an Initial Searching Point

Each possible solution of the TS is a point in the feasible space of the optimization problem. In an N-dimensional optimization problem, a solution point is an array of size $1 \times N$. The TS starts with a single solution of size $1 \times N$:

$$Point = X = \left(x_1, x_2, \ldots, x_i, \ldots, x_N\right) \tag{6.1}$$

where X = a solution of the optimization problem, x_i = ith decision variable of the solution X, and N = number of decision variables. The decision variable values $(x_1, x_2, x_3, \ldots, x_N)$ are represented as a predefined set of values for discrete problems. The optimization algorithm starts with an initial point known as the searching point that is randomly selected from the discrete decision space when solving combinatorial problems (see Section 2.6). Neighboring points (new solutions) are considered near the searching point.

6.4 Neighboring Points

The TS was developed to solve combinatorial (finite solution set) optimization problems. The decision variables take discrete values in this instance. The neighborhood of a solution is made of all the solutions in which the value of one decision variable is changed to its immediate adjacent values in a sorted list of discrete values. Assume that Figure 6.2 portrays the decision space of an optimization problem with a two-dimensional discrete decision space. In Figure 6.2 the circles denote possible answers for the problem.

Suppose that in Figure 6.2, the solution (3,2), which is encircled by a square, is selected as searching point. All the possible solutions that are connected

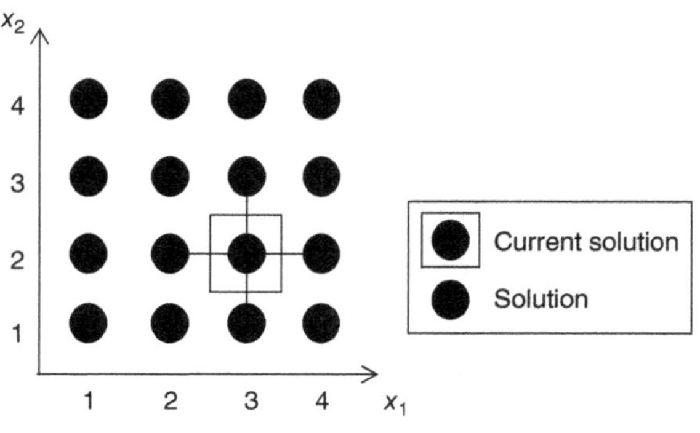

Figure 6.2 The decision space of a two-dimensional optimization problem.

with lines to (3,2) are neighboring solutions. In an N-dimensional space if $X=(x_1,x_2,\ldots,x_i,\ldots,x_N)$ is the searching point, the number of neighbor solutions surrounding the searching point is $2N$, which are represented as follows:

$$M=\begin{bmatrix} X_1^{(new)}=\left(x_1',x_2,\ldots,x_i,\ldots,x_N\right) \\ X_2^{(new)}=\left(x_1,x_2',\ldots,x_i,\ldots,x_N\right) \\ \vdots \\ X_i^{(new)}=\left(x_1,x_2,\ldots,x_i',\ldots,x_N\right) \\ \vdots \\ X_N^{(new)}=\left(x_1,x_2,\ldots,x_i,\ldots,x_N'\right) \\ X_{N+1}^{(new)}=\left(x_1'',x_2,\ldots,x_i,\ldots,x_N\right) \\ X_{N+2}^{(new)}=\left(x_1,x_2'',\ldots,x_i,\ldots,x_N\right) \\ \vdots \\ X_{N+i}^{(new)}=\left(x_1,x_2,\ldots,x_i'',\ldots,x_N\right) \\ \vdots \\ X_{2N}^{(new)}=\left(x_1,x_2,\ldots,x_i,\ldots,x_N''\right) \end{bmatrix} \quad (6.2)$$

$$x_i'=x_i+\varepsilon,\quad i=1,2,\ldots,N \quad (6.3)$$

$$x_i''=x_i-\varepsilon,\quad i=1,2,\ldots,N \quad (6.4)$$

in which M = the matrix of the neighbor points around X, $X_i^{(new)}$ = ith neighboring solution near the searching point X, x_i = the value of decision variable ith of the searching point, x_i' and x_i'' = the new values of decision variable ith, and ε = the length of steps in a discrete decision space.

The search movements are generalized as follows:

$$X^{(new)}=X\pm\left(\varepsilon\cdot e_{1\times N}\right) \quad (6.5)$$

in which $X^{(new)}$ = a neighbor solution around X, e = a specified direction vector that is a matrix of zeroes and ones, and ε = the length of steps in a discrete decision space.

For specific combinatorial problem at hand, various alternative possible neighborhood structures can be used according to the search space. Choosing a search space and a neighborhood structure has a key role in the design of any TS algorithm.

6.5 Tabu Lists

The TS features principles of artificial intelligence because it acts based on memory structures that prevent repetitive movements, thus improving the search for optima within the decision space of the optimization problem more effectively. The TS prevents moving back to previously visited solutions using memories recorded in a tabu list that memorize the history of the search. The tabu list is a key feature of the TS, which is a deterministic method and allows movements to worse neighboring solutions if there is no improving solution. It is probable then that the TS algorithm might fall into an iterative cycle without making any progress toward an optimal solution. Cycling defines a situation whereby the TS algorithm is trapped into repetitive movements. When cycling occurs there must be intervention to break it up. The tabu list is used to prevent cycling.

Tabus can be defined and memorized in different ways. One possibility is recording a solution that is visited. Moving to a previously visited solution is forbidden. Memorizing the visited solution is the most efficient method to prevent cycling, yet it is seldom used because this requires a lot of storage making it expensive to evaluate whether or not a potential move is tabu. On the other hand, another commonly used tabu is memorizing the transformation performed on the current solution and prohibiting reverse transformations. In this manner following a move from solution X to X' precludes returning from X' to X, but arriving at X through another route is allowed.

It is recommended that transformation and reverse transformation be recorded as tabu. Recalling Figure 6.2, if the current searching point is (3,2) and the TS algorithm moves to point (2,2) as new searching point, moving back from (2,2) to (3,2) is forbidden and considered as a tabu. The TS algorithm cannot follow a previously traveled path. Yet, it is possible to return to point (3,2). Imagine that after two iterations, the search reaches point (3,3). The algorithm may move back to point (3,2) through another route that is not a tabu. This would introduce cycling because the algorithm will again move from (3,2) to (2,2), which can be solved by recording the forward transformation and the reverse transformation as tabu. In the previously mentioned example, the algorithm moves from the current searching point (3,2) to point (2,2) requires recording the moves (3,2) to (2,2) and (2,2) to (3,2) as tabu. This would eliminate cycling involving these moves.

6.6 Updating the Tabu List

Tabus are memorized for a predefined number of iterations and then they are removed from the tabu list. Generally, the basic purpose of a tabu list is to avoid returning to a previously visited point (solution). The probability of cycling caused by following a sequence of moves that ends with a previously visited solution is inversely related to the distance of the current searching point from that previous solution (Glover, 1989).

The number of iterations during which tabu lists are saved is called the tabu tenure (δ), which is a user-defined parameter of the algorithm. In this manner the tabu list keeps information on the last δ moves (iterations) that have been performed during the search process. For example, a solution that becomes a tabu at iteration 1 with the tabu tenure equal to 3 ($\delta = 3$) is deleted from the tabu list at iteration 4. When a new solution (information) is added to a full tabu list, it results in the removal of the oldest solution or information from the list.

Standard tabu lists usually have a fixed length. Yet, tabu lists of variable lengths have been implemented (Glover, 1989, 1990; Skorin-Kapov, 1990; Taillard, 1990, 1991).

6.7 Attributive Memory

Many types of memories for TS have been developed during past years to enhance its capacity. One of the most common of such memories is attributive memory. This type of memory records information about the properties or attributes of solutions. The most prevalent attributive memory approaches are frequency-based memory and recency-based memory.

6.7.1 Frequency-Based Memory

Frequency-based memory keeps information that facilitates selecting feature moves. Specifically, the TS algorithm saves the number of times that each solution has been selected to be the searching point, and this information is called frequency-based memory, which excises a penalty that is proportional to the frequency with which a solution is visited by subtracting a penalty value from the fitness value of the solution as follows:

$$F'(X) = F(X) - \mu \tag{6.6}$$

in which $F'(X)$ = penalized fitness function of the solution X, $F(X)$ = fitness function of the solution X, and μ = number of times that solution X has been visited. Equation (6.6) applies to a maximization problem.

The desirability of a solution is evaluated by its penalized fitness value rather than its fitness function value. Thus, frequency-based memory would select between two neighboring solutions that have the same fitness value and with the lower frequency of visitation (under maximization).

6.7.2 Recency-Based Memory

Two possible methods to record tabus were cited earlier, one that saves a solution that is visited and the other that saves the search movements that are performed.

There is another way of recording tabus whereby attributes of solutions are memorized as a tabu list. This type of tabu list is called recency-based memory, the most prevalent memory structure applied in TS implementations. This memory structure keeps track of solutions' attributes that have changed during the recent past. For example, recalling Figure 6.2, suppose that the search starts from point (3,2) and moves to point (2,2). It is clear that in moving from (3,2) to (2,2), the value of decision variable x_1 is reduced, and by moving back from (2,2) to (3,2), increasing the value of x_1 is necessary. Moving back to the previous solution and cycling does not occur if increasing the value of x_1 is forbidden. Increasing the value of x_1 is therefore considered as a tabu for the δ next iterations. The selected attributes or properties of solutions recently visited are labeled tabu active in recency-based memory. Solutions that have tabu-active elements, or particular combinations of these attributes, are those that become tabu.

A tabu list that is constructed on recency-based memory is easy to save and to read information from. Selected attributes occur in solutions recently visited. If these attributes are labeled tabu active, then solutions that have never been visited but share the same tabu-active attributes are prevented from being visited by the search algorithm. To illustrate, suppose the search starts with point (3,2) in Figure 6.3 and moves to (2,2). Assume that δ (tabu tenure) is equal to 2. This movement decreases the value of x_1, which precludes increasing the value of x_1 in the next two iterations. The TS algorithm can thus move to (2,1), (2,3), or (1,2), but it cannot move to (3,2). Suppose that point (2,3) is selected to be the next searching point. Decreasing the value of x_2 would be a tabu in the next movement. In the third iteration the searching point is (2,3) and the two neighboring points (2,4) and (1,3) are non-tabu points, whereas the two neighboring points (2,2) and (3,3) are tabu (see Figure 6.3). It is evident in Figure 6.3 that the point (3,3) has not been visited during the search, but it is considered as a tabu because increasing the value of x_1 from 2 to 3 is tabu. Thus, it is necessary to cancel tabus in some instances. The action by which the algorithm ignores tabus is called aspiration criteria.

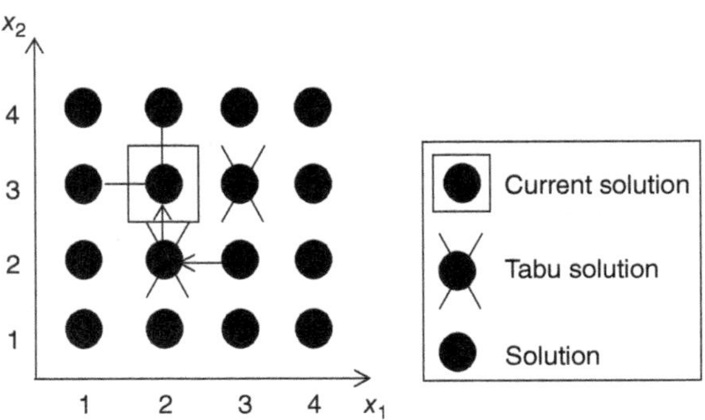

Figure 6.3 Illustration of steps based on recency-based memory in a two-dimensional decision space.

6.8 Aspiration Criteria

Tabus may prohibit rewarding moves even when there is no danger of cycling, or they may lead to an overall stagnation of the searching process. Consequently, the aspiration criterion frees the algorithm to move to a tabu point under certain favorable conditions. A commonly used aspiration criterion is a condition in which a tabu movement leads to a better solution than the best solution that has been encountered in previous moves.

6.9 Intensification and Diversification Strategies

Intensification and diversification strategies are applied in some versions of TS. Intensification strategies have been commonly implemented based on long-term memory so that a set of elite solutions are selected, their components serve to construct new neighbors, and diversification strategies encourage examining unvisited regions by generating solutions significantly different from those searched earlier (Glover and Laguna, 1997).

These strategies ensure that all areas of the search space are adequately explored. A frequency-based tabu list keeps track of the search area. The frequency index of a previously visited solution is increased whenever it is revisited. The diversification strategy set a threshold of the frequency index. A value of the frequency index larger than a predefined threshold implies that a region has been explored frequently and the search process is restarted with a new, randomly generated point. This diversification strategy is similar to the restart mechanism of other stochastic optimization approaches. The restart in the TS, however, is guided by historical information based on an intensification strategy.

6.10 Termination Criteria

The termination criterion determines when to terminate the TS algorithm. Selecting a good termination criterion has an important role for the correct convergence of the algorithm. The number of algorithmic iterations, the amount of improvement of the objective function between consecutive iterations, and the run time are common termination criteria for the TS algorithm.

6.11 User-Defined Parameters of the TS

The value of the tabu tenure (δ) and the criterion used to decide when to terminate the algorithm are the user-defined parameters of the TS. Choosing an appropriate δ is important. Large values of δ make the algorithm to move gradually away from points visited during the previous δ iterations. A small value for δ provides more mobility to the algorithm. Previous experience with the TS reveals that there is a stable range of δ that prevents cycling and leads to good solutions. A good choice of the parameters is related to the decision space of a particular problem. In general, optimal parameters for one problem may not function well for other problems. Consequently, determining a good parameter setting often requires the execution of a large number of time-consuming experiments. The setting of suitable parameters relies principally on in-depth experience with the type of problem being solved. Yet, a reasonable method for finding appropriate values of the TS parameters is performing sensitivity analysis whereby combinations of parameters are tested with multiple runs for each combination. From these results the analyst may gain clear guidance on the choice of parameters that produce near-optimal solutions.

6.12 Pseudocode of the TS

```
Begin
  Input parameters of the algorithm and initial data
  Let X* = the best point and X = the current search
    point
  Generate a search point (X) randomly and evaluate
    its fitness value
  Set X* = X
  While (termination criteria are not satisfied)
     Generate neighbor points around the searching
       point and evaluate their fitness values
     If all neighbor points are tabus and cannot satisfy
       the aspiration criteria
          Stop the algorithm and report the best point (X*)
     End if
     Select the best neighbor point which is not tabu
       or satisfies the aspiration criteria
     Put X = the selected point
     If X is better than X*
          Set X* = X
     End if
     Update the tabu list
  End while
  Report the best point (X*)
End
```

6.13 Conclusion

This chapter described the TS, which is a meta-heuristic algorithm for combinatorial optimization. First, a brief literature review of development and applications of the TS was presented. This was followed by a description of the fundamentals of the TS and its algorithmic nuances. A pseudocode of the TS closed this chapter.

References

Bland, J. A. (1995). "Discrete-variable optimal structural design using tabu search." Structural Optimization, 10(2), 87–93.

Fanni, A., Giacinto, G., Marchesi, M., and Serri, A. (1999). "Tabu search coupled with deterministic strategies for the optimal design of MRI devices." International Journal of Applied Electromagnetics and Mechanics, 10(1), 21–31.

Glover, F. (1986). "Future paths for integer programming and links to artificial intelligence." Computers and Operations Research, 13, 533–549.

Glover, F. (1989). "Tabu search: Part I." ORSA Journal on Computing, 1(3), 190–206.

Glover, F. (1990). "Tabu search: Part II." ORSA Journal on Computing, 2(1), 4–32.

Glover, F. and Laguna, M. (1997). "Tabu search." Kluwer Academic, Norwell, MA.

Haghighi, A. and Bakhshipour, A. (2014). "Deterministic integrated optimization model for sewage collection networks using tabu search." Journal of Water Resources Planning and Management, 141(1), 04014045.

Hajji, M., Fares, A., Glover, F., and Driss, O. (2010). "Water pump scheduling system using scatter search, tabu search and neural networks: The case of bouregreg water system in Morocco." In: Palmer, R. N. (Ed.), World environmental and water resources congress 2010: Challenges of change, May 16–20, American Society of Civil Engineers (ASCE), Reston, VA, 822–832.

Hajji, O., Brisset, S., and Brochet, P. (2004). "A new tabu search method for optimization with continuous parameters." IEEE Transactions on Magnetics, 40(2), 1184–1187.

Martinez, S., Merwade, V., and Maidment, D. (2010). "Linking GIS, hydraulic modeling, and tabu search for optimizing a water level-monitoring network in south Florida." Journal of Water Resources Planning and Management, 136(2), 167–176.

Misevicius, A. (2005). "A tabu search algorithm for the quadratic assignment problem." Computational Optimization and Applications, 30(1), 95–111.

Nara, K., Hayashi, Y., Ikeda, K., and Ashizawa, T. (2001). "Application of tabu search to optimal placement of distributed generators." IEEE Power Engineering Society Winter Meeting, Columbus, OH, January 27–31, Piscataway, NJ: Institute of Electrical and Electronics Engineers (IEEE).

Nourzad, S. and Afshar, M. (2009). "Industrial distribution system simulation for optimal water resource assignment using probabilistic tabu search." World Environmental and Water Resources Congress 2009, Kansas City, MO, May 17–21, Reston, VA: American Society of Civil Engineers (ASCE), 1–10.

Skorin-Kapov, J. (1990). "Tabu search applied to the quadratic assignment problem." ORSA Journal on Computing, 2(1), 33–45.

Taillard, É. (1990). "Some efficient heuristic methods for the flow shop sequencing problem." European Journal of Operational Research, 47(1), 65–74.

Taillard, É. (1991). "Robust taboo search for the quadratic assignment problem." Parallel Computing, 17(4–5), 443–455.

Wang, C., Quan, H., and Xu, X. (1999). "Optimal design of multiproduct batch chemical process using tabu search." Computers and Chemical Engineering, 23(3), 427–437.

Yeh, S. F., Chang, Y. J., and Lin, M. D. (2013). "Optimal design of sewer network by tabu search and simulated annealing." 2013 IEEE International Conference on Industrial Engineering and Engineering Management, Bangkok, Thailand, December 10–13, Piscataway, NJ: Institute of Electrical and Electronics Engineers (IEEE).

7

Ant Colony Optimization

Summary

This chapter describes ant colony optimization (ACO). The basic concepts of the ACO are derived from analogy to the foraging behavior of ants. The chapter begins with a brief literature review highlighting the development and applications of the ACO. This is followed by a description of the ACO's algorithm. A pseudocode of the ACO closes the chapter.

7.1 Introduction

Ant colony optimization (ACO) was introduced by Dorigo et al. (1991, 1996). It attempts to simulate in algorithmic fashion the foraging behavior of ants. Several varieties of ACO algorithms have appeared since its original inception, and those include the elitist ant system (AS) (Dorigo, 1992; Dorigo et al., 1996), Ant-Q (Gambardella and Dorigo, 1995), ant colony system (Gambardella and Dorigo, 1996; Dorigo and Gambardella, 1997), max–min AS (Stutzle and Hoos, 2000), and the hypercube AS (Blum and Dorigo, 2004). The ACO has solved various types of problems such as vehicle routing (Reimann et al., 2004), project scheduling (Merkle et al., 2002), and open shop scheduling (Blum, 2005). Various types of ant-based algorithms have found frequent implementations in civil engineering and structural optimization (Christodoulou, 2010; Lee, 2012; Sharafi et al., 2012). Abadi and Jalili (2006) applied the ACO for network vulnerability analysis. Effatnejad et al. (2013) implemented the ACO for determining the feasible optimal solution of economic dispatching. Afshar et al. (2015) wrote a state-of-the-art review of the ACO's applications to water resources management.

Meta-Heuristic and Evolutionary Algorithms for Engineering Optimization,
First Edition. Omid Bozorg-Haddad, Mohammad Solgi, and Hugo A. Loáiciga.

7.2 Mapping Ant Colony Optimization (ACO) to Ants' Foraging Behavior

The ACO takes inspiration from the foraging behavior of some ant species that deposit pheromone on the ground to mark favorable paths for colony members to follow to procure food. Many ant species take advantage of a particular type of communication called stigmergy in which workers are stimulated by the performance they have achieved. Stigmergy is differentiated from other types of communication in two distinct ways. First, stigmergy is an indirect, non-symbolic form of communication mediated by the environment. Ants exchange information by modifying their environment. Second, stigmergic information is local. Therefore, the information exchanged by stigmergy is only available for those who see the location where it is released.

This type of communication explains the strategy followed by ants to collect and transport food to their nest. Many ant species release a substance called pheromone along their tracks while they are walking to or from a food source. Other ants detect the presence of pheromone and follow paths where the pheromone concentration is present. This tactic allows ants to transport food to their nest in a remarkably effective way.

The pheromone-laying and pheromone-following behavior of ants was studied by Deneubourg et al. (1990). In a double-bridge experiment, the nest of an ant colony was connected to a food source using two bridges that were equal in length. Ants could reach the food crossing either one of the bridges. The results of experiment demonstrated that initially the bridges were selected arbitrarily by the ants. Due to random fluctuations one of the two bridges exhibited a higher concentration of pheromone than the other after some time, and, therefore, more ants moved through the former bridge. This brought a further amount of pheromone to that bridge, making it a more attractive route, and eventually the whole colony converged toward using the same bridge.

The pheromone-laying mechanism helps ants find the shortest path between a food source and their nest. When one of the bridges is shorter, ants that move through the shorter path reach the food sooner and increase the concentration of pheromone on the way back. Goss et al. (1989) considered a variant of the double-bridge experiment in which one bridge is significantly longer than the other. In this case, the stochastic fluctuations in the initial choice of a bridge are much reduced, and a second mechanism plays an important role. Specifically, the ants choosing by chance the short bridge are the first to reach the nest. The short bridge receives larger amounts of pheromone earlier than the long one, and this fact increases the probability that more ants select it for transport to and from a food source. Figure 7.1 shows a schematic of the double-bridge experiment.

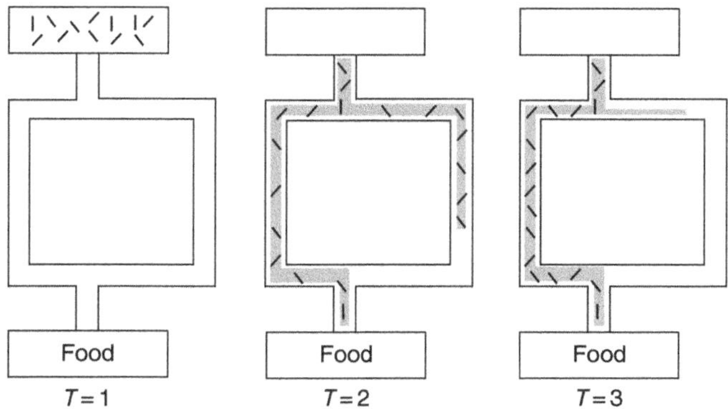

Figure 7.1 A double-bridge experiment with pathways of unequal length.

Goss et al. (1989) presented a model to evaluate probability of an ant to select the first bridge (ϕ_1) where μ_1 ants have selected the first bridge and μ_2 ants have selected the other bridge in a double-bridge experiment as follows:

$$\phi_1 = \frac{(\mu_1 + z)^h}{(\mu_2 + z)^h + (\mu_1 + z)^h} \tag{7.1}$$

in which ϕ_1 = the probability of an ant selecting the first bridge, μ_1 = the number of ants that have selected the first bridge, μ_2 = the number of ants that have selected the second bridge, and z and h = parameters that must be fitted to the experimental data.

The ants represent solutions in ACO. The path of an ant is a set of decision variables that constitute a solution of the optimization problem in ACO. In other words, the tour of an ant from nest to food represents a possible solution of the optimization problem. Each ant has a fitness value corresponding to the value of the objective function of the optimization problem that reflects the length of its tour. The better the fitness value, the shorter the length of the tour. Each ant leaves a concentration of pheromone in a specific area of the decision space according to its fitness value that marks its path. New ants (solutions) are made based on information left by previous ants in the decision space. Table 7.1 lists the characteristics of the ACO.

The ACO starts by generating a set of random solutions made up of decision variables that are selected from a predefined set of discrete values. The fitness values of all the solutions are evaluated. Then, proportionate to the fitness values of solutions, concentrations of pheromone are assigned to the decision space. The concentration of pheromone shows desirability. The parts of the decision space that make fitter solutions achieve more

Table 7.1 The characteristics of the ACO.

General algorithm (see Section 2.13)	Ant colony optimization
Decision variable	Track path of ant
Solution	Ant
Old solution	Old ant
New solution	New ant
Best solution	–
Fitness function	Pheromone
Initial solution	Random ant
Selection	–
Process of generating new solutions	Based-information stochastic mechanism

concentration of pheromone. The sum of the pheromone of a specific value of a decision variable is equal to all the pheromone left by all ants that possess that value. New ants (solutions) are constructed in the next algorithmic step based on information retrieved from previous ants. New solutions are generated randomly with a stochastic function that assigns a probability to allowable values of each decision variable according to its pheromone. Values that have higher concentration of pheromone are more likely to be selected. Concentrations of pheromone are added to the decision space to generate new solutions after evaluating the fitness values of newly generated solutions if the termination criteria are not satisfied. Otherwise the algorithm ends. Figure 7.2 depicts the flowchart of the ACO.

7.3 Creating an Initial Population

An ant's track in any dimension of an N-dimensional space represents a decision variable of the optimization problem. An ant is known as an array of size $1 \times N$ that describes the ant's path. This array is defined as follows:

$$Ant = X = \left(x_1, x_2, \ldots, x_i, \ldots, x_N\right) \tag{7.2}$$

where X = a solution of the optimization problem, x_i = ith decision variable of solution X, and N = number of decision variables. The decision variable values $(x_1, x_2, x_3, \ldots, x_N)$ are chosen from a set of predefined values for discrete problems. The ACO solves problems with discrete domain; each decision variable i takes a value from a predefined set of values V_i as follows:

$$V_i = \left\{v_{i,1}, v_{i,2}, \ldots, v_{i,d}, \ldots, v_{i,D_i}\right\}, \quad i = 1,2,\ldots,N \tag{7.3}$$

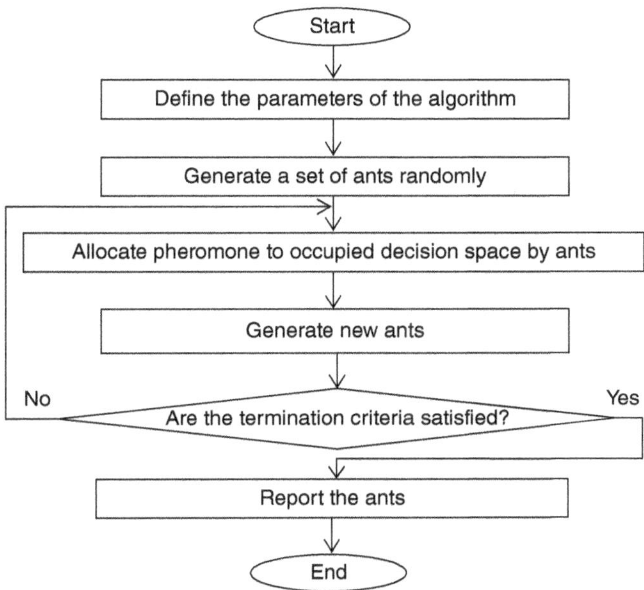

Figure 7.2 The flowchart of the ACO.

in which V_i = the set of predefined values for ith decision variable, $v_{i,d}$ = dth possible value for the ith decision variable, and D_i = total number of possible values for the ith decision variable.

A solution is represented as a graph that connects decision variables to each other defining a pathway or solution as shown in Figure 7.3. The number of layers of the graph equals the number of decision variables, and the number of nodes in a particular layer equals the number of discrete probable values permitted for the corresponding design variable. Thus, each node on the graph is associated with a permissible discrete value of a design variable.

The ACO algorithm begins with randomly generating a matrix (see Section 2.6) of size $M \times N$ (where M and N denote the size of the population of solutions and the number of decision variables, respectively). Hence, the matrix of solutions that is generated randomly is as follows (there are M rows or solutions, and each solution contains N decision variable):

$$Population = \begin{bmatrix} X_1 \\ X_2 \\ \vdots \\ X_j \\ \vdots \\ X_M \end{bmatrix} = \begin{bmatrix} x_{1,1} & x_{1,2} & \cdots & x_{1,i} & \cdots & x_{1,N} \\ x_{2,1} & x_{2,2} & \cdots & x_{2,i} & \cdots & x_{2,N} \\ & & & \vdots & & \\ x_{j,1} & x_{j,2} & \cdots & x_{j,i} & \cdots & x_{j,N} \\ & & & \vdots & & \\ x_{M,1} & x_{M,2} & \cdots & x_{M,i} & \cdots & x_{M,N} \end{bmatrix} \tag{7.4}$$

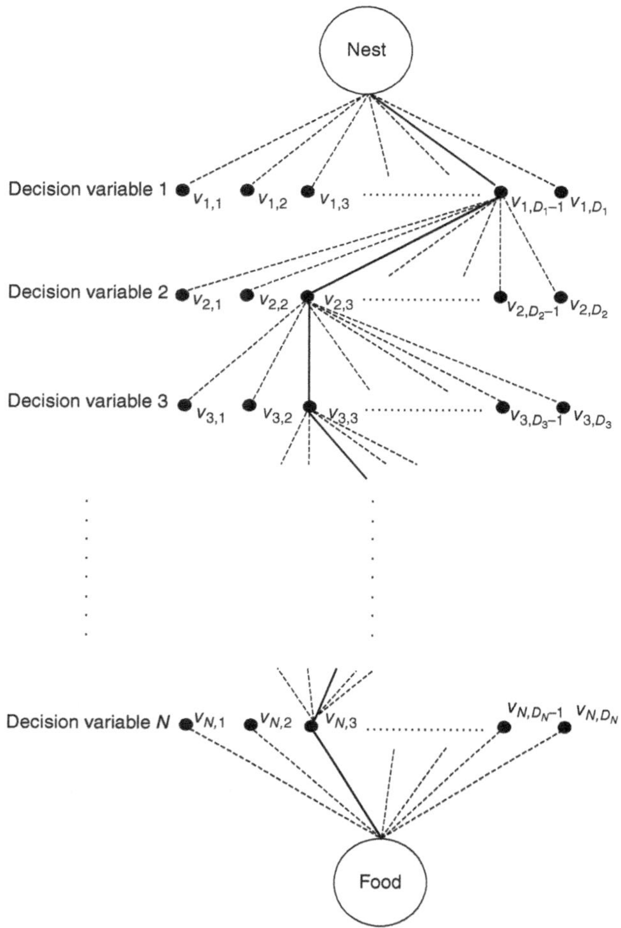

Figure 7.3 Representation of a solution in the ACO; there are *M* such solutions.

in which $X_j = j$th solution, $x_{j,i} = i$th decision variable of the jth solution, and M = population size (the number of solutions). The value of $x_{j,i}$ is randomly selected from a set V_i (see Equation (7.3)).

7.4 Allocating Pheromone to the Decision Space

In contrast to most other meta-heuristic and evolutionary algorithms, the ACO allocates desirability to the decision space instead of the solutions to find the best region of the decision space or the best mixture of decision variables. Most meta-heuristic and evolutionary algorithms generate new solutions using old solutions, say, by mixing old solutions or by generating new solutions in the neighborhood of old solutions. The ACO searches different points of the decision space and adds information to the decision space. New solutions are randomly constructed based on the information in the decision space. The ACO allocates a concentration of pheromone to each decision variable's value according to the fitness value of a solution. The more fit a solution is, the higher the pheromone concentration, and vice versa. In other words, values that make better solutions achieve higher concentration of pheromone in comparison with values that make worse solutions.

N arrays of size $1 \times D_i$ are employed to allocate pheromone to the decision space so that each of them is assigned to one decision variable as follows:

$$C_i = \left(c_{i,1}, c_{i,2}, \ldots, c_{i,d}, \ldots, c_{i,D_i}\right), \quad i = 1,2,\ldots,N \tag{7.5}$$

in which C_i = pheromone matrix for the ith decision variable and $c_{i,d}$ = pheromone concentration of the dth possible value of the ith decision variable.

The elements of the matrix C equal zero at the beginning of the algorithmic optimization. They are updated during the algorithmic search, taking positive values. The aim of the pheromone update is to increase the pheromone concentration of good or promising decision variable's values. The pheromone allocation is achieved by (1) decreasing all the pheromone values through pheromone evaporation and (2) increasing the pheromone levels associated with a chosen set of good solutions. Solutions (ants) are generated and their fitness values are evaluated. The concentration of pheromone for the dth possible value of the ith decision variable is updated as follows:

$$c_{i,d}^{(new)} = \left(1 - \rho\right) \times c_{i,d} + \sum_{j=1}^{M} \Delta c_{i,d}^{(j)}, \quad d = 1,2,\ldots,D_i, \quad i = 1,2,\ldots,N \tag{7.6}$$

in which $c_{i,d}^{(new)}$ = new concentration of pheromone of the dth possible value of the decision variable, ρ = evaporation rate, and $\Delta c_{i,d}^{(j)}$ = the quantity of pheromone laid on the dth possible value of the ith decision variable by the jth ant. The value of $\Delta c_{i,d}^{(j)}$ corresponds to the fitness value of the jth solution, and it is estimated as follows in a minimization problem:

$$\Delta c_{i,d}^{j} = \begin{cases} \dfrac{Q}{F(X_j)} & \text{if } x_{j,i} = v_{i,d} \\ 0 & \text{if Otherwise} \end{cases}, \quad j = 1,2,\ldots,M, \quad i = 1,2,\ldots,N, \quad d = 1,2,\ldots,D_i \tag{7.7}$$

in which Q = a constant value and $F(X_j)$ = fitness value of the jth solution.

Equation (7.7) was proposed by Dorigo et al. (1996) to solve the salesman problem in which the fitness value must be a positive number; otherwise it is not acceptable for estimating the concentration of pheromone. It follows from Equation (7.7) that in a minimization problem, the concentration of pheromone is inversely proportional to the fitness value. By extension it follows that in a maximization problem, the concentration of pheromone is proportional to the fitness value. The concentration of pheromone can be zero or a positive value. The ACO solves optimization problems relying on a formula that relates the pheromone concentrations and fitness values according to Equation (7.7).

New solutions are generated after evaluating of concentration of pheromone for possible values for all decision variables so that the values with higher concentration of pheromone have a better chance of being selected for new solutions.

7.5 Generation of New Solutions

New solutions are generated through a stochastic process. Each decision variable i is assigned a value with a probability that depends on the concentration of pheromone. Specifically, a probability $P_{i,d}$ is assigned to each possible value d of decision variable i as follows:

$$P_{i,d} = \frac{(c_{i,d})^{\alpha} \times (\eta_{i,d})^{\beta}}{\sum_{k=1}^{D_i}\left[(c_{i,k})^{\alpha} \times (\eta_{i,k})^{\beta}\right]}, \quad d=1,2,\ldots,D_i, \quad i=1,2,\ldots,N \tag{7.8}$$

in which $P_{i,d}$ = probability that the dth possible value ($v_{i,d}$) be selected for the ith decision variable, $\eta_{i,d}$ = a heuristic value for the dth possible value *of* the ith decision variable, and α and β = parameters that control the relative importance of the pheromone versus the heuristic information ($\eta_{i,d}$). The heuristic information shows the desirability of selecting possible values that help the algorithmic search of the decision space more efficiently. For instance, in the structural design problem presented in Chapter 1, the purpose is minimizing the weight of the structure. Before solving the problem it is clear that a smaller cross-sectional area would produce a lighter structure. It is possible to assign a heuristic value to the cross-sectional area so that less cross-sectional area has a larger heuristic value and, thus, a high probability of being chosen. The choice of the parameters α and β renders this feature optional. The sum of the probabilities of the possible values of each decision variable is equal to one:

$$\sum_{d=1}^{D_i} P_{i,d} = 1, \quad i=1,2,\ldots,N \tag{7.9}$$

The values of the decision variables of a new solution are randomly selected based on the evaluated probabilities. To accomplish this one first calculate a cumulative probability for all the possible values of each decision variable as follows:

$$\psi_{i,r} = \sum_{d=1}^{r} P_{i,d}, \quad i = 1,2,\ldots,N, \quad r = 1,2,\ldots,D_i \tag{7.10}$$

in which $\psi_{i,r}$ = cumulative probability of the rth possible value of the ith decision variable.

Thereafter, a random number (*Rand*) in the range [0,1] is generated. If *Rand* is less than $\psi_{i,1}$, the first value ($v_{i,1}$) is selected; otherwise the rth value is selected such that *Rand* is greater than $\psi_{i,r-1}$ and less or equal than $\psi_{i,r}$ ($\psi_{i,r-1} < Rand \leq \psi_{i,r}$). This procedure assigns randomly a value from the set V_i to each decision variable i of solution j. If component $v_{i,d}$ of set $V_i = \{v_{i,1}, v_{i,2}, \ldots, v_{i,d}, \ldots, v_{i,D_i}\}$ is assigned to the ith decision variable of the jth new solution, we have

$$x'_{j,i} = v_{i,d}, \quad i = 1,2,\ldots,N, \quad j = 1,2,\ldots,M \tag{7.11}$$

in which $x'_{j,i}$ = new value of the ith decision variable of the jth new solution.

The new solutions are constructed after evaluation of all the decision variables as follows:

$$X_j^{(new)} = \left(x'_{j,1}, x'_{j,2}, \ldots, x'_{j,i}, \ldots, x'_{j,N}\right), \quad j = 1,2,\ldots,M \tag{7.12}$$

in which $X_j^{(new)}$ = new solution j. The M newly generated solutions displace all of the old solutions.

7.6 Termination Criteria

The termination criterion determines when to end the algorithm. Selecting a good termination criterion is essential for the correct convergence of the ACO algorithm. The number of iterations, the incremental improvement of the objective function between consecutive iterations, and the run time are common convergence criteria for the ACO.

7.7 User-Defined Parameters of the ACO

The size of the population of solutions (M), the evaporation rate (ρ), the control parameters of pheromone (α), heuristic information (β), and the termination criterion of the algorithm are user-defined parameters of the ACO. A good choice of the parameters is related to the decision space of a particular problem, and usually the optimal parameter setting for one problem is of limited utility for any other problem. Consequently, determining a good set of parameters requires the execution of a large number of time-consuming experiments.

The setting of suitable parameters relies principally on in-depth experience with the type of problem being solved. Yet, a reasonable method for finding appropriate values of the ACO parameters is performing sensitivity analysis whereby combinations of parameters are tested with multiple runs for each combination. From these results the analyst may gain clear guidance on the choice of parameters that produce near-optimal solutions.

7.8 Pseudocode of the ACO

```
Begin
  Input parameters of the algorithm and initial data
  Let M=population size and N=number of decision
    variables
  Let Di=number of possible values for decision
    variable i
  Generate M initial possible solutions randomly
  While (termination criteria are not satisfied)
     Evaluate fitness values for all solutions
     For i=1 to N
         For d=1 to Di
             Update pheromone concentration of possible
               value d for decision variable i
             Evaluate probability of possible value d
               to be selected
         Next d
     Next i
     For j=1 to M
         For i=1 to N
             Randomly select a value for decision
               variable i among possible values based
               on their probabilities
         Next i
     Next j
  End while
  Report the ants or solutions
End
```

7.9 Conclusion

This chapter described ACO. It included a brief literature review of the ACO, a mathematical statement of its algorithm, and a pseudocode.

References

Abadi, M. and Jalili, S. (2006). "An ant colony optimization algorithm for network vulnerability analysis." Iranian Journal of Electrical and Electronic Engineering, 2(3,4), 106–120.

Afshar, A., Massoumi, F., Afshar, A., and Mariño, M. A. (2015). "State of the art review of ant colony optimization applications in water resource management." Water Resources Management, 29(11), 3891–3904.

Blum, C. (2005). "Beam-ACO-hybridizing ant colony optimization with beam search: An application to open shop scheduling." Computers and Operations Research, 32(6), 1565–1591.

Blum, C. and Dorigo, M. (2004). "The hyper-cube framework for ant colony optimization." IEEE Transactions on Systems, Man, and Cybernetics–Part B, 34(2), 1161–1172.

Christodoulou, S. (2010). "Scheduling resource-constrained projects with ant colony optimization artificial agents." Journal of Computing in Civil Engineering, 24(1), 45–55.

Deneubourg, J. L., Aron, S., Goss, S., and Pasteels, J. M. (1990). "The self-organizing exploratory pattern of the Argentine ant." Journal of Insect Behavior, 3(2), 159–168.

Dorigo, M. (1992). "Optimization, learning, and natural algorithms." PhD Dissertation, Politecnico di Milano, Milan.

Dorigo, M. and Gambardella, L. M. (1997). "Ant colonies for the traveling salesman problem." BioSystems, 43(2), 73–81.

Dorigo, M., Maniezzo, V., and Colorni, A. (1991). "Positive feedback as a search strategy." Dipartimento di Elettronica, Politecnico di Milano, Milano, Technical Report No: 91-016.

Dorigo, M., Maniezzo, V., and Colorni, A. (1996). "The ant system: Optimization by a colony of cooperating ants." Institute of Electrical and Electronics Engineers (IEEE), Transactions on Systems Man and Cybernetics, Part B, 26(1), 29–42.

Effatnejad, R., Aliyari, H., Tadayyoni, H., and Abdollahshirazi, A. (2013). "Novel optimization based on the ant colony for economic dispatch." International Journal on Technical and Physical Problems of Engineering, 5(2), 75–80.

Gambardella, L. M. and Dorigo, M. (1995). "Ant-Q: A reinforcement learning approach to the traveling salesman problem." In: Prieditis, A. and Russell, S. (Eds.), Proceedings of the twelfth international conference on machine learning (ML-95), Tahoe City, CA, July 9–12, Morgan Kaufmann, Palo Alto, CA.

Gambardella, L. M. and Dorigo, M. (1996). "Solving symmetric and asymmetric TSPs by ant colonies." Proceedings of the 1996 IEEE International Conference on Evolutionary Computation (ICEC'96), Nagoya University, Japan, May 20–22, Piscataway, NJ: Institute of Electrical and Electronics Engineers (IEEE), 622–627.

Goss, S., Aron, S., Deneubourg, J. L., and Pasteels, J. M. (1989). "Self-organized shortcuts in the Argentine ant." Naturwissenschaften, 76(12), 579–581.

Lee, H. (2012). "Integrating simulation and ant colony optimization to improve the service facility layout in a station." Journal of Computing in Civil Engineering, 26(2), 259–269.

Merkle, D., Middendorf, M., and Schmeck, H. (2002). "Ant colony optimization for resource-constrained project scheduling." IEEE Transactions on Evolutionary Computation, 6(4), 333–346.

Reimann, M., Doerner, K., and Hartl, R. F. (2004). "D-ants: Savings based ants divide and conquer the vehicle routing problems." Computers and Operations Research, 31(4), 563–591.

Sharafi, P., Hadi, M. N. S., and Teh, L. H. (2012). "Heuristic approach for optimum cost and layout design of 3D reinforced concrete frames." Journal of Structural Engineering, 138(7), 853–863.

Stutzle, T. and Hoos, H. H. (2000). "Max–min ant system." Future Generation Computer Systems, 16(8), 889–914.

8

Particle Swarm Optimization

Summary

This chapter describes the particle swarm optimization (PSO) technique, which is inspired by the swarming strategies of various organisms in nature. The next section reviews a few implementations of the PSO. The remainder of this chapter describes the PSO algorithm and presents a pseudocode for its implementation.

8.1 Introduction

Kennedy and Eberhart (1995) developed the particle swarm optimization (PSO) algorithm as a meta-heuristic algorithm based on the social behavior exhibited by birds or fishes when striving to reach a destination. Balci and Valenzuela (2004) presented a technique that uses the PSO combined with the Lagrangian relaxation (LR) framework to solve a power generator scheduling problem known as the unit commitment problem. Chuanwen and Bompard (2005) applied a self-adaptive chaotic PSO algorithm for optimal hydroelectric plant dispatch model based on the rule of maximizing the benefit in a deregulated environment. The proposed approach introduced chaos mapping, and the self-adaptive chaotic PSO algorithm increased the mapping convergence rate and associated precision. Suribabu and Neelakantan (2006) used the Environmental Protection Agency's hydraulic network simulator (EPANET) and the PSO algorithm in a combined simulation and optimization model to design a water distribution pipeline network. Matott et al. (2006) identified the PSO algorithm as an effective technique for solving pump-and-treat optimization problems with analytic element flow models. Izquierdo et al. (2008) applied the PSO algorithm to the optimization of a wastewater collection network. Results showed that the algorithm's performance and the calculated results were consistent with those calculated with dynamic programming to solve the same problem under the same conditions. Fallah-Mehdipour et al. (2011) proposed a multi-objective PSO

Meta-Heuristic and Evolutionary Algorithms for Engineering Optimization, First Edition. Omid Bozorg-Haddad, Mohammad Solgi, and Hugo A. Loáiciga.

(MOPSO) approach in multipurpose multi-reservoir operation. MOPSO calculated near-optimal Pareto fronts. Ostadrahimi et al. (2012) employed a multi-swarm PSO to extract multi-reservoir operation rules and showed that the PSO algorithm outperformed the common implicit stochastic optimization approach in the real-time operation of a reservoir system. Gaur et al. (2012) coupled artificial neural network (ANN) and the PSO for the management of the Dore river basin in France. An analytic element method (AEM)-based flow model was applied to generate the data set for the training and testing of the ANN. A coupled ANN-PSO was implemented to minimize the pumping cost of wells and compared with AEM-PSO. ANN-PSO reduced the computational burden significantly in the analysis of various management scenarios. Kumar and Reddy (2012) employed elitist-mutated PSO (EMPSO) as an efficient and reliable swarm intelligence-based approach in multipurpose reservoir operation. Results demonstrated that the EMPSO performed better than the PSO algorithm. Noory et al. (2012) presented a discrete PSO algorithm for optimizing irrigation water allocation and multi-crop planning. Saadatpour and Afshar (2013) implemented the MOPSO in a pollution spill response management model in reservoirs. They coupled CE-QUAL-W2 with the MOPSO algorithm to obtain a desirable near-optimal reservoir operation strategy and/or emergency planning in a selective withdrawal framework. Fallah-Mehdipour et al. (2013) calculated multi-crop planning rules in a reservoir system with the PSO algorithm, the genetic algorithm (GA), and shuffled frog leaping algorithm (SFLA). They maximized the total net benefit of the water resources system by supplying irrigation water for a proposed multi-cropping pattern over the planning horizon. Qu and Lou (2013) proposed a PSO algorithm based on the immune evolutionary algorithm (IEA) to optimal allocation of regional water resources. The results of the survey demonstrated that the performance of the presented algorithm to solve the issue of optimal allocation of regional water resources is reliable and reasonable. Bozorg-Haddad et al. (2013) compared the performance of the PSO algorithm with that of the pattern search (PS) algorithm for calibration of numerical groundwater models. Orouji et al. (2014) proposed a hybrid algorithm, linking the PSO and SFLA, to solve the resource-constrained project scheduling problem (RCPSP). The RCPSP minimized the duration of a construction project considering resource limitations and the timing of activities. The hybrid PSO-SFLA proved more capable in determining an optimal solution with fewer iterations compared with the individual application of the PSO and SFLA.

8.2 Mapping Particle Swarm Optimization (PSO) to the Social Behavior of Some Animals

This chapter deals with the application of computational algorithms to biological–social systems and, more specifically, to the collective behaviors of individuals interacting with their environment and each other. These systems are known as swarm intelligence. Swarm intelligence defines the discipline that deals with natural and artificial systems consisting of many individuals who coordinate among themselves using decentralized control and self-organization. Swarm intelligence focuses on the collective behavior resulted from the local interactions between individuals and between individuals and their environment. Beehives, ant colonies, fish schools, bird flocks, and animal herds are examples of systems with swarm intelligence. The PSO is one of the most common examples of swarm intelligence.

The PSO algorithm is based on the social behavior of birds. It simulates the behaviors of bird flocks. Suppose that a group of birds are randomly looking for food in an area. Imagine that seeker birds do not know where the food is. One effective strategy to find food is for birds to follow the bird that is known to be nearest to the food. The PSO acts according to the previous example and employs a numerical analog to solve optimization problems.

The PSO designates each single solution in the decision space of the optimization problem as a bird and is called a particle. All the particles have fitness values that are evaluated by the objective function to be optimized, which measures their distances to food. Each particle also has velocity that directs the flying of the particle. The best particle is the leader, and other particles follow the leader. The particles fly through the decision space of the problem by following the leader. Each particle determines its next position based on (1) its best individual position so far occupied and (2) the best position achieved in the group. In other words, each particle is updated by two positions. The first one is the best position that the particle has occupied so far. The other is the best position achieved so far by any particle in the population of particles. The first position is the best individual position, and the second one is the best global position. Table 8.1 lists the characteristics of the PSO.

The PSO starts with the position and velocity of particles randomly initialized within the search space (see Section 2.6). The fitness values of the particles are calculated. These first fitness values and positions are the best individual fitness values and the best individual positions, respectively. The best position among all particles is the global best position. The position and velocity of each particle are updated to generate new solutions based on their personal and global best positions. In the next iteration the fitness values of the updated particles are recalculated, and the personal and global best positions are updated. In this manner the new particles' positions and velocities are generated. The PSO algorithm continues updating the individual and global best positions and generating new positions until the termination criteria are met. Figure 8.1 shows the flowchart of the PSO.

Table 8.1 The characteristics of the PSO.

General algorithm (see Section 2.13)	Particle swarm optimization
Decision variable	Particle's position in each dimension
Solution	Particle's position
Old solution	Old position of particle
New solution	New position of particle
Best solution	Leader of particles
Fitness function	Distance between particle and food
Initial solution	Random particle
Selection	–
Process of generating new solution	Flying with a specific velocity and direction

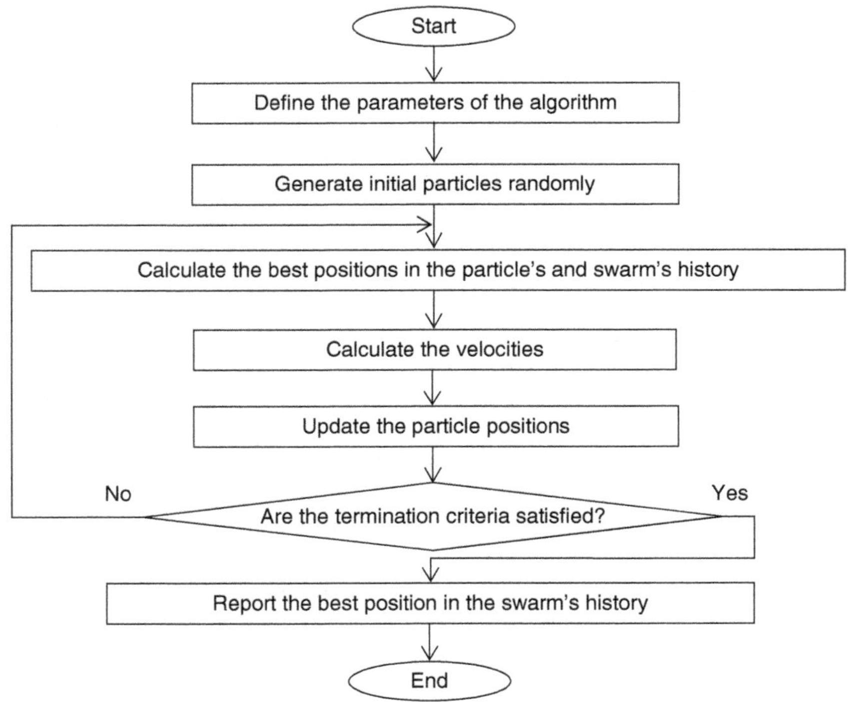

Figure 8.1 The flowchart of the PSO.

8.3 Creating an Initial Population of Particles

The PSO designates each possible solution of the optimization problem as a particle. The particles' positions represent the decision variables in an N-dimensional optimization problem. Particles are specified as an array of size $1 \times N$. This array is defined as follows:

$$Particle = X = \left(x_1, x_2, \ldots, x_i, \ldots, x_N\right) \tag{8.1}$$

where X = a possible solution of the optimization problem, x_i = ith decision variable of solution X, and N = number of decision variables. The decision variable values $(x_1, x_2, x_3, \ldots, x_N)$ can be represented as floating point number (real values) or as a predefined set for continuous and discrete problems, respectively.

The PSO algorithm starts by randomly generating a matrix of particles (see Section 2.6) of size $M \times N$ (where M and N denote the size of the population of solutions and the number of decision variables, respectively). Hence, the matrix of solutions is as follows (there are M rows or solutions; each solution contains N decision variables):

$$Population = \begin{bmatrix} X_1 \\ X_2 \\ \vdots \\ X_j \\ \vdots \\ X_M \end{bmatrix} = \begin{bmatrix} x_{1,1} & x_{1,2} & \cdots & x_{1,i} & \cdots & x_{1,N} \\ x_{2,1} & x_{2,2} & \cdots & x_{2,i} & \cdots & x_{2,N} \\ & & & \vdots & & \\ x_{j,1} & x_{j,2} & \cdots & x_{j,i} & \cdots & x_{j,N} \\ & & & \vdots & & \\ x_{M,1} & x_{M,2} & \cdots & x_{M,i} & \cdots & x_{M,N} \end{bmatrix} \tag{8.2}$$

in which X_j = jth solution, $x_{j,i}$ = ith decision variable of the jth solution, and M = population size.

8.4 The Individual and Global Best Positions

Each particle moves through the decision space based on the individual and global best positions. Each particle attempts to achieve the best or optimal position in the decision space with two types of parameters (*Pbest* and *Gbest*). *Pbest* and *Gbest* are the best positions in the particle's and swarm's histories, respectively. For each particle j, there is a *Pbest* as follows:

$$Pbest_j = \left(p_{j,1}, p_{j,2}, \ldots, p_{j,i}, \ldots, p_{j,N}\right), \quad j = 1, 2, \ldots, M \tag{8.3}$$

in which $Pbest_j$ = the best position of the jth particle and $p_{j,i}$ = the best position of the jth particle in the ith dimension.

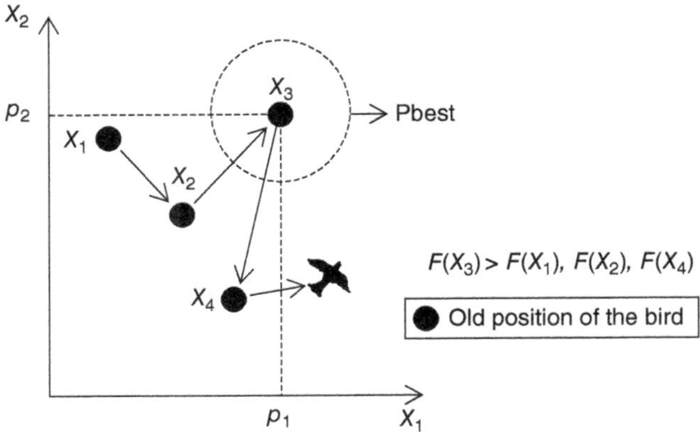

Figure 8.2 Concepts of the best individual position in a two-dimensional maximization problem.

Gbest is an array $1 \times N$ whose elements define the best position achieved in the swarm:

$$Gbest = \left(g_1, g_2, \ldots, g_i, \ldots, g_N\right), \quad all\, j \tag{8.4}$$

where *Gbest* = the best position in the swarm's history and g_i = the best position in the swarm's history in the *i*th dimension.

The initial population of particles are generated randomly (see Section 2.6), and their fitness values are calculated. These first fitness values and positions constitute the best individual fitness values and the best individual positions (*Pbest*). The best position among all particles is considered as the global best position (*Gbest*). In each iteration of the algorithm, *Pbest* and *Gbest* are updated. Each particle's best individual position (*Pbest*) is updated if the fitness value of the particle's new position is better than *Pbest*. Figure 8.2 illustrates the concept of the best individual position in a maximization problem. In Figure 8.2 the route of a bird flying on a two-dimensional space is depicted. This bird experiences different positions including X_1, X_2, X_3, and X_4. This bird finds different amounts of food [$F(X)$] at each position so that position X_3 is the best among all. The bird memorizes the position X_3 as the *Pbest*. It memorizes this position as the *Pbest* until it reaches a position with more food.

The concept of the global best position in a maximization problem is shown in Figure 8.3. *Gbest* is the best point that is calculated during the optimization search.

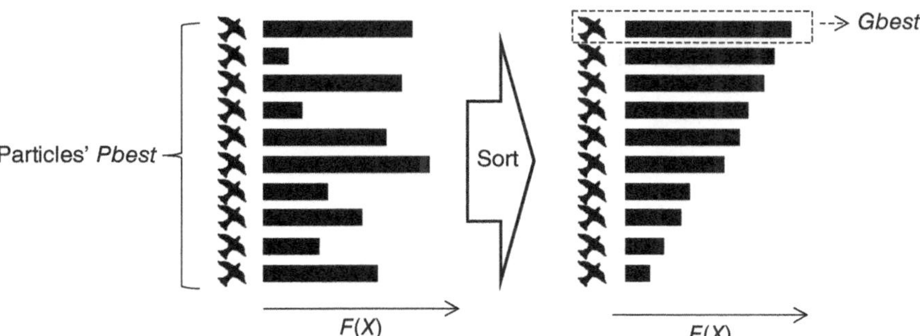

Figure 8.3 Concept of the global best position in a maximization problem.

8.5 Velocities of Particles

Each particle has a velocity that directs the flying of the particle and determines its next position of the particle. The particles' velocities are used to update their positions. The velocity is calculated based on *Gbest* and *Pbest*. The velocity of the *j*th particle (V_j) represented by an $1 \times N$ array is as follows:

$$V_j = \left(v_{j,1}, v_{j,2}, \ldots, v_{j,i}, \ldots, v_{j,N}\right), \quad j = 1,2,\ldots,M \tag{8.5}$$

in which V_j = the velocity of the *j*th particle and $v_{j,i}$ = the velocity of the *j*th particle in the *i*th dimension that is calculated as follows:

$$v_{j,i}^{(new)} = \omega \times v_{j,i} + C_1 \times Rand \times \left(p_{j,i} - x_{j,i}\right) + C_2 \times Rand \times \left(g_i - x_{j,i}\right) \\ j = 1,2,\ldots,M, \quad i = 1,2,\ldots,N \tag{8.6}$$

in which $v_{j,i}^{(new)}$ = the new velocity of the *j*th particle in the *i*th dimension, $v_{j,i}$ = the previous velocity of the *j*th particle in the *i*th dimension, ω = inertia weight parameter, *Rand* = a random value in the range [0,1], C_1 = cognitive parameter, and C_2 = social parameter (C_1 and C_2 control the movement of *Pbest* and *Gbest* toward an optimal point and usually $C_1 = C_2 = 2$). Movement along different directions toward *Gbest* and *Pbest* is possible if C_1 and C_2 are larger than one.

Lower and upper bounds for velocity limit the variation of a particle's velocity as follows:

$$V_i^{(L)} \le v_{j,i}^{(new)} \le V_i^{(U)}, \quad j = 1,2,\ldots,M, \quad i = 1,2,\ldots,N \tag{8.7}$$

in which $V_i^{(L)}$ and $V_i^{(U)}$ = the lower and upper bound of the velocity along the *i*th dimension, respectively.

The inertia weight parameter ω has an important role in swarm convergence and controls the effects of the current velocity. Large or small values of ω cause searching in a wide or narrow space, respectively. The inertia weight parameter may change as the algorithm progresses as follows:

$$\omega_t = \omega_0 - \left[\left(\omega_0 - \omega_T\right) \times \frac{t}{T} \right], \quad t = 1,2,\ldots,T \tag{8.8}$$

in which ω_0 = initial inertia weight, ω_T = inertia weight for the last iteration, and T = total number of iterations. The value of ω changes through the iterations.

8.6 Updating the Positions of Particles

Particles move to new positions from the old position to generate new positions. This movement of particles is performed based on the velocities. A particle's position is updated as follows:

$$X_j^{(new)} = \left(x'_{j,1}, x'_{j,2}, \ldots, x'_{j,i}, \ldots, x'_{j,N}\right), \quad j = 1,2,\ldots,M \tag{8.9}$$

$$x'_{j,i} = x_{j,i} + v_{j,i}^{(new)}, \quad j = 1,2,\ldots,M, \quad i = 1,2,\ldots,N \tag{8.10}$$

in which $X_j^{(new)}$ = jth new solution and $x'_{j,i}$ = new value of ith decision variable of the jth new solution. The M newly generated solutions displace all the old solutions.

8.7 Termination Criteria

The termination criteria determine when to terminate the algorithm. Selecting a good termination criterion has an important role to correct the convergence of the algorithm. The number of iterations, the run time, and the improvement of the solution between consecutive iterations are common termination criteria for the PSO algorithm.

8.8 User-Defined Parameters of the PSO

The size of the population of solutions (M), the value of the initial inertia weight (ω_0), the final value of the inertia weight (ω_T), and the termination criterion are user-defined parameters of the PSO. A good choice of the parameters is related to the decision space of a particular problem, and usually the optimal parameter setting for one problem is of limited utility for other problems. Consequently, determining a good parameter setting often requires the execution of numerous time-consuming experiments. In-depth practice and experience are keys to choosing appropriate algorithmic parameters. However, a reasonable method for finding appropriate values for the parameters is performing sensitivity analysis. Combinations of parameters are applied to run the algorithm. The algorithm is run several times for each combination of parameters. This captures the variability of the solution due to the randomness of the PSO algorithm. By comparing the solutions across the combination of parameters, the analyst is guided to choose appropriate values.

8.9 Pseudocode of the PSO

```
Begin
  Input parameters of the algorithm and initial data
  Generate M initial possible solutions and evaluate
    their fitness values
  Initialize the velocities of all the solutions
    randomly
  For j = 1 to M
      Put Pbest_j = solution j
  Next j
  Set Gbest = the best solution in the population
  While (termination criteria are not satisfied)
      For j = 1 to M
          If the fitness value of solution j is better
            than that of Pbest_j
              Put Pbest_j = solution j
          End if
      Next j
      If the best solution is better than Gbest
          Substitute the best solution for Gbest
      End if
      For j = 1 to M
          Calculate the velocity of solution j
          Control the velocity of solution j
          Update solution j
      Next j
      Evaluate fitness value for all solutions
      Update inertia weight parameter (ω)
  End while
  Report the best solution
End
```

8.10 Conclusion

This chapter described the PSO, which is based on the strategies of swarms in their search for food. The chapter presented a brief literature review of the PSO, its algorithmic fundamentals, and a pseudocode.

References

Balci, H. H. and Valenzuela, J. F. (2004). “Scheduling electric power generators using particle swarm optimization combined with the Lagrangian relaxation method.” International Journal of Applied Mathematics and Computer Science, 14(3), 411–421.

Bozorg-Haddad, O., Tabari, M. M. R., Fallah-Mehdipour, E., and Mariño, M. A. (2013). “Groundwater model calibration by meta-heuristic algorithms.” Water Resources Management, 27(7), 2515–2529.

Chuanwen, J. and Bompard, E. (2005). “A self-adaptive chaotic particle swarm algorithm for short term hydroelectric system scheduling in deregulated environment.” Energy Conversion and Management, 46(17), 2689–2696.

Fallah-Mehdipour, E., Bozorg-Haddad, O., and Mariño, M. A. (2011). “MOPSO algorithm and its application in multipurpose multireservoir operation.” Journal of Hydroinformatics, 13(4), 794–811.

Fallah-Mehdipour, E., Bozorg-Haddad, O., and Mariño, M. A. (2013). “Extraction of multi-crop planning rules in a reservoir system: Application of evolutionary algorithms.” Journal of Irrigation and Drainage Engineering, 139(6), 490–498.

Gaur, S., Sudheer, Ch., Graillot, D., Chahar, B. R., and Kumar, D. N. (2012). “Application of artificial neural networks and particle swarm optimization for the management of groundwater resources.” Water Resources Management, 27(3), 927–941.

Izquierdo, J., Montalvo, I., Perez, R., and Fuertes, V. (2008). “Design optimization of wastewater collection network by PSO.” Computers and Mathematics with Applications, 56(3), 777–784.

Kennedy, J. and Eberhart, R. (1995). “Particle swarm optimization.” Proceeding of International Conference on Neural Networks, Perth, Australia, November 27 to December 1, Piscataway, NJ: Institute of Electrical and Electronics Engineers (IEEE), 1942–1948.

Kumar, N. and Reddy, M. J. (2012). “Multipurpose reservoir operation using particle swarm optimization.” Journal of Water Resources Planning and Management, 133(3), 192–201.

Matott, L. S., Rabideau, A. J., and Craig, J. R. (2006). “Pump-and treat optimization using analytic element method flow models.” Advances in Water Resources, 29(5), 760–775.

Noory, H., Liaghat, A., Parsinejad, M., and Bozorg-Haddad, O. (2012). "Optimizing irrigation water allocation and multicrop planning using discrete PSO algorithm." Journal of Irrigation and Drainage Engineering, 138(5), 437–444.

Orouji, H., Bozorg-Haddad, O., Fallah-Mehdipour, E., and Mariño, M. (2014). "Extraction of decision alternatives in project management: Application of hybrid PSO-SFLA." Journal of Management in Engineering, 30(1), 50–59.

Ostadrahimi, L., Mariño, M. A., and Afshar, A. (2012). "Multi-reservoir operation rules: Multi-swarm PSO-based optimization approach." Water Resources Management, 26(2), 407–427.

Qu, G. and Lou, Z. (2013). "Application of particle swarm algorithm in the optimal allocation of regional water resources based on immune evolutionary algorithm." Journal of Shanghai Jiaotong University (Science), 18(5), 634–640.

Saadatpour, M. and Afshar, A. (2013). "Multi objective simulation-optimization approach in pollution spill response management model in reservoirs." Water Resources Management, 27(6), 1851–1865.

Suribabu, C. R. and Neelakantan, T. R. (2006). "Design of water distribution networks using particle swarm optimization." Urban Water Journal, 3(2), 111–120.

9

Differential Evolution

Summary

This chapter describes differential evolution (DE), which is a parallel direct search method that takes advantage of some features of evolutionary algorithms (EAs). The DE is a simple yet powerful meta-heuristic method. This chapter begins with a brief literature review about the DE and its applications, followed by a presentation of the DE's fundamentals and a pseudocode.

9.1 Introduction

Differential evolution (DE) was developed by Storn and Price (1997). The DE was designed primarily for continuous optimization problems. Lampinen and Zelinka (1999) presented a modified DE for discrete optimization. Vesterstrom and Thomsen (2004) demonstrated that DE had a better performance in comparison with other optimization techniques such as the genetic algorithm (GA) and particle swarm optimization (PSO). The DE algorithm has been successfully applied to solve a wide range of optimization problems such as clustering, pattern recognition, and neural network training (Price et al., 2005). Tang et al. (2008) applied the DE to structural system identification. Lakshminarasimman and Subramanian (2008) implemented the DE for optimization of power systems. Qing (2009) demonstrated different applications of the DE in electrical engineering. Wang et al. (2009) applied the DE for optimum design of truss structures. Gong et al. (2009) applied the DE to optimal engineering design. Xu et al. (2012) implemented the DE to estimate parameter of a nonlinear Muskingum model applied for flood prediction in water resources management. Gonuguntla et al. (2015) presented a modified DE with adaptive parameter specification.

Meta-Heuristic and Evolutionary Algorithms for Engineering Optimization,
First Edition. Omid Bozorg-Haddad, Mohammad Solgi, and Hugo A. Loáiciga.

9.2 Differential Evolution (DE) Fundamentals

DE is a parallel direct search method that takes advantage of some features of evolutionary algorithms (EAs). In other words, the DE is a method that optimizes a problem by iteratively improving candidate solutions with respect to a given measure of quality.

Direct search methods can solve a variety of numerical problems with emphasis on the use of simple strategies rather than complex tactics, which makes them well suited for computational processing. The phrase "direct search" refers to sequential examination of trial solutions. Direct search methods compare each trial solution with the best solution previously obtained, and the result of the comparison determines the next trial solution. Direct search techniques employ straightforward search strategies. These techniques have some features that distinguish them from classical methods. They have solved problems that defied classical methods. They have calculated solutions for some problems faster than classical methods. In addition, direct search techniques apply repeated identical arithmetic operations with a simple logic that are easily coded for computer processing (Hooke and Jeeves, 1961).

Direct search methods choose a point B randomly that is called the base point. A second point, P1, is chosen, and if it is better than B, then it replaces the base point; otherwise, B is not changed. This process continues by comparing each new point with the current base point. The "strategy" for selecting new trial points is determined by a set of "states" that constitute the memory of the search process. The number of states is finite. There are an arbitrary initial state and a final state that stops the search. The other states represent various situations that arise as a function of the results of the trials made. The strategy implemented to select new points is dictated by various aspects of the problem, including the structure of its decision space. The strategy includes the choice of an initial base point, the rules of transition between states, and the rules for selecting trial points as a function of the current state and the base point. Direct search designates a trial point as a move or step from the base point. The move is a success if the trial point is better than the base point and is a failure otherwise. The states make up part of the algorithmic logic influencing moves in the same general direction as those that have recently succeeded. The states suggest new directions if recent moves have failed; and they decide when no further progress can be made. A diagnosis that no further progress can be made does not always indicate that the solution to an optimization problem has been found.

The DE algorithm was designed by using the common concepts of EAs, such as multipoint searching, use of crossover, mutation, and selection operators, and it has some unique characteristics that make it different from many other EAs. The major differences with EAs are how offspring are

Table 9.1 The characteristics of the DE.

General algorithm (see Section 2.13)	Differential evolution
Decision variable	Coordinate of agent's position
Solution	Agent (position)
Old solution	Old agent
New solution	Trial agent
Best solution	–
Fitness function	Desirability of the agent
Initial solution	Random agent
Selection	Greedy criterion
Process of generating new solution	Mutation and crossover

generated and in the selection mechanism that the DE applies to transition from one generation of solutions to the next.

The DE algorithm has a population of candidate solutions that are called agents. The components of each agent in an N-dimensional space constitute the decision variables of the optimization problem. These agents are moved in the decision space by using crossover and mutation operators that combine and change their positions. In other words, trial agents are produced based on old agents. In this respect the DE resembles EAs because it applies genetic operators to produce new solutions. Selection of solutions is done based on greedy criteria. If the new position of an agent is an improvement, it is accepted and it replaces the old solution. An improved trial agent is known as a success and is added to the population of solutions. Otherwise the trial agent is a failure, and it is discarded from the direct search. Table 9.1 lists the characteristics of the DE.

The DE starts by randomly generating a set of solutions (see Section 2.6) known as the initial population of agents. A new trial solution or agent is generated for each agent (solution). The generation of a trial agent requires that three agents from the old population be randomly selected and a new solution be generated using a heuristic function. This process is known as mutation. A crossover operator is implemented to combine the old agent and the newly generated solution. This produces a trial solution. The trial solution replaces the old solution if it has a better fitness value according to the greedy criteria. Otherwise, the old solution remains in the population. Trial solutions are again generated if the termination criteria are not satisfied. Otherwise, the final population is reported, and the algorithm ends. Figure 9.1 shows the flowchart of the DE.

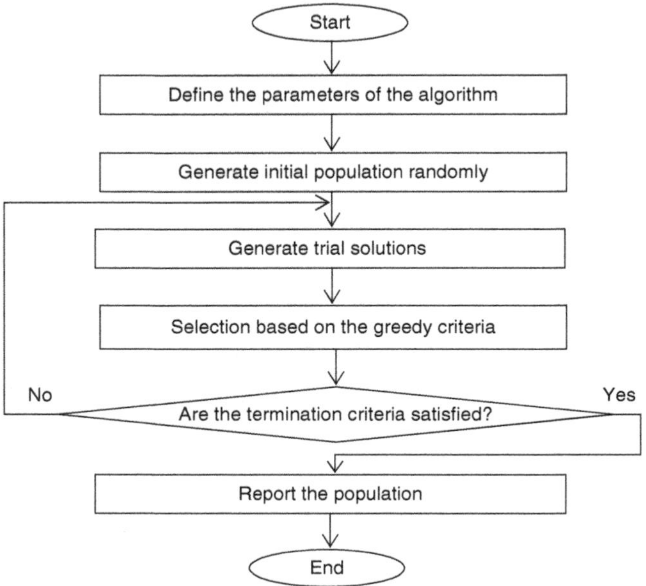

Figure 9.1 The flowchart of the DE.

9.3 Creating an Initial Population

Each possible solution of the optimization problem found by DE is called an agent. In an N-dimensional optimization problem, an agent's component in any dimension of the N-dimensional space is a decision variable of the optimization problem. An agent is an array of size $1 \times N$:

$$Agent = X = (x_1, x_2, \ldots, x_i, \ldots, x_N) \tag{9.1}$$

where X = a possible or tentative solution of the optimization problem, x_i = ith decision variable of solution X, and N = number of decision variables. A matrix of agents of size $M \times N$ is randomly generated (see Section 2.6), where M and N are the size of population and the number of decision variables, respectively, to start the optimization algorithm. This matrix is represented as follows:

$$Population = \begin{bmatrix} X_1 \\ X_2 \\ \vdots \\ X_j \\ \vdots \\ X_M \end{bmatrix} = \begin{bmatrix} x_{1,1} & x_{1,2} & \cdots & x_{1,i} & \cdots & x_{1,N} \\ x_{2,1} & x_{2,2} & \cdots & x_{2,i} & \cdots & x_{2,N} \\ & & & \vdots & & \\ x_{j,1} & x_{j,2} & \cdots & x_{j,i} & \cdots & x_{j,N} \\ & & & \vdots & & \\ x_{M,1} & x_{M,2} & \cdots & x_{M,i} & \cdots & x_{M,N} \end{bmatrix} \tag{9.2}$$

in which $X_j = j$th solution, $x_{j,i} = i$th decision variable of the jth solution, and M = population size. Each of the decision variable values of the jth solution $(x_{j,1}, x_{j,2}, x_{j,3}, \ldots, x_{j,N})$ is represented as a floating point number (real values). The DE solves problems with continuous decision space.

9.4 Generating Trial Solutions

The DE generates new trial solutions by means of mutation and crossover operations acting in series. A new trial solution X'_j is first generated by mutation. The crossover operator is applied to the newly generated solution that combines the old solution and the newly generated solution to produce a trial solution $X_j^{(new)}$. This process is performed on each member of the population of solutions.

9.4.1 Mutation

Mutation produces a new solution that is mixed with an old solution by crossover to generate a trial solution. Mutation is executed by selecting randomly three solutions—X_r, $X_{r'}$, and $X_{r''}$—from the present population. Thereafter, new solutions are generated as follows:

$$X'_j = \left(x'_{j,1}, x'_{j,2}, \ldots, x'_{j,i}, \ldots, x'_{j,N}\right), \quad j = 1,2,\ldots,M \tag{9.3}$$

$$x'_{j,i} = x_{r,i} + \delta \times \left(x_{r',i} - x_{r'',i}\right), \quad i = 1,2,\ldots,N, \quad j = 1,2,\ldots,M, \quad r \neq r' \neq r'' \neq j \tag{9.4}$$

$$r = Irnd(1, M), \quad r \neq r' \neq r'' \neq j \tag{9.5}$$

$$r' = Irnd(1, M), \quad r \neq r' \neq r'' \neq j \tag{9.6}$$

$$r'' = Irnd(1, M), \quad r \neq r' \neq r'' \neq j \tag{9.7}$$

in which X'_j = new jth mutated solution; $x'_{j,i} = i$th decision variable of the jth new mutated solution; δ = mutation factor that is a value in the range [0,2]; $x_{r,i}$, $x_{r',i}$, $x_{r'',i} = i$th decision variable of the rth, r'th, and r''th solutions, respectively; and $Irnd(1,M)$ = an integer random number in the range [1,M]; r, r', and, r'' = counters of randomly chosen solutions that are different from each other and from the counter j that designates the target old solution. A new solution generated by mutation is crossed over with the target old solution to generate a trial solution.

9.4.2 Crossover

Crossover combines the targeted old solution (X_j) with the newly generated solution (X'_j) to generate a trial solution $X_j^{(new)}$ as follows:

$$X_j^{(new)} = \left(x''_{j,1}, x''_{j,2}, \ldots, x''_{j,i}, \ldots, x''_{j,N}\right) \tag{9.8}$$

$$x''_{j,i} = \begin{cases} x'_{j,i} & \text{if } Rand \le C \text{ or } i = b_j \\ x_{j,i} & \text{if } Rand > C \text{ and } i \ne b_j \end{cases}, \quad i = 1,2,\ldots,N, \quad j = 1,2,\ldots,M \tag{9.9}$$

$$b_j = Irnd(1,N), \quad j = 1,2,\ldots,M \tag{9.10}$$

in which $X_j^{(new)}$ = jth trial solution, $x''_{j,i}$ = ith decision variable of the jth trial solution, $Rand$ = a random number in the range [0,1], C = crossover constant that is a predefined value in the range [0,1], and b_j = a randomly chosen index that denotes a decision variable of the jth solution and ensures that $X_j^{(new)}$ has at least one decision variable from X'_j; otherwise no new solution is produced.

9.5 Greedy Criteria

The trial solution $X_j^{(new)}$ is compared with the old solution X_j to determine whether or not it becomes a member of the population of solutions. If the trial solution has better fitness value than the old solution, it replaces it. Otherwise, the trial solution is deleted and the old solution is kept in the population. This selection process is called greedy criteria.

9.6 Termination Criteria

The termination criterion determines when to terminate the algorithm. Selecting a good termination criterion has an important role on the correct convergence of the DE algorithm. The number of iterations, the magnitude of the improvement of the solution between consecutive iterations, and the run time are common convergence criteria for the DE.

9.7 User-Defined Parameters of the DE

The population size (M), the mutation factor (δ), the crossover constant (C), and the termination criteria are user-defined parameters of the DE. A good choice of the parameters depends on the decision space of a particular problem, and the optimal parameter setting for one problem is of limited utility for other problems. Consequently, determining a good parameter setting often requires the execution of a large number of computational experiments. A reasonable method for finding appropriate values for the parameters is performing sensitivity analysis, whereby combinations of parameters are tested and the algorithm is run several times for each combination to account for the random nature of the solution algorithm.

In this manner the analyst obtains a distribution of solutions and associated objective function values for each combination of parameters. A comparison of the results from all the combination of parameters provides guidance on a proper choice of the algorithmic parameters.

9.8 Pseudocode of the DE

```
Begin
  Input parameters of the algorithm and initial data
  Generate M initial possible solutions
  Evaluate fitness value for solutions
  While (termination criteria are not satisfied)
     For j = 1 to M
          Generate solution X'_j by mutation
          Generate trial solution X_j^(new) by crossover
            between X'_j and old solution X_j
          Evaluate fitness value of trial solution X_j^(new)
          If trial solution X_j^(new) is better than old
            solution X_j
              Put X_j = X_j^(new)
          End if
     Next j
  End while
  Report the population of solutions
End
```

9.9 Conclusion

This chapter described the DE, which is a parallel direct search method that takes advantage of some features of EAs. The chapter presented the algorithmic fundamentals of the DE and a pseudocode.

References

Gong, W., Cai, Z., and Zhu, L. (2009). "An efficient multi-objective differential evolution algorithm for engineering design." Structural and Multidisciplinary Optimization, 4(2), 137–157.

Gonuguntla, V., Mallipeddi, R., and Veluvolu, K. C. (2015). "Differential evolution with population and strategy parameter adaptation." Mathematical Problems in Engineering, 2015, 287607.

Hooke, R. and Jeeves, T. A. (1961). "Direct search solution of numerical and statistical problems." Journal of the ACM, 8(2), 212–229.

Lakshminarasimman, L. and Subramanian, S. (2008). "Applications of differential evolution in power system optimization." In: Kacprzyk, J. and Chakraborty, U. K. (Eds.), Advances in differential evolution, Studies in computational intelligence, Vol. 143, Springer, Berlin, Heidelberg, 257–273.

Lampinen, J. and Zelinka, I. (1999). "Mechanical engineering design optimization by differential evolution." In: Come, D., Dorigo, M., and Glover, F. (Eds.), New ideas in optimization, McGraw Hill, Maidenhead, 127–146.

Price, K. V., Storn, R. M., and Lampinen, J. A. (2005). "Differential evolution: A practical approach to global optimization." Springer-Verlag, Berlin, Heidelberg.

Qing, A. (2009). "Differential evolution: Fundamentals and applications in electrical engineering." John Wiley & Sons, Inc., Hoboken, NJ.

Storn, R. and Price, K. (1997). "Differential evolution-a simple and efficient heuristic for global optimization over continuous spaces." Journal of Global Optimization, 11(4), 341–359.

Tang, H., Xue, S., and Fan, C. (2008). "Differential evolution strategy for structural system identification." Computers and Structures, 86(21–22), 2004–2012.

Vesterstrom, J. and Thomsen, R. A. (2004). "Comparative study of differential evolution, particle swarm optimization, and evolutionary algorithms on numerical benchmark problems." Congress on Evolutionary Computation (CEC2004), Portland, OR, June 19–23, Piscataway, NJ: Institute of Electrical and Electronics Engineers (IEEE), 1980–1987.

Wang, Z., Tang, H., and Li, P. (2009). "Optimum design of truss structures based on differential evolution strategy." 2009 International Conference on Information Engineering and Computer Science, Wuhan, China, December 19–20, Piscataway, NJ: Institute of Electrical and Electronics Engineers (IEEE).

Xu, D. M., Qiu, L., and Chen, S. Y. (2012). "Estimation of nonlinear Muskingum model parameter using differential evolution." Journal of Hydrologic Engineering, 17(2), 348–353.

10

Harmony Search

Summary

This chapter describes harmony search (HS), which is a meta-heuristic algorithm for discrete optimization. A brief literature review of the HS is presented, followed by a description of its algorithmic fundamentals. A pseudocode of the HS closes this chapter.

10.1 Introduction

Geem et al. (2001) developed harmony search (HS) inspired by the harmony found in many musical compositions. The HS has been applied to various benchmarking and real-world optimization problems. Kim et al. (2001) implemented the HS for estimation of the nonlinear Muskingum model for flood routing. Geem et al. (2002) applied the HS to find optimal design of water distribution networks. Lee and Geem (2004) implemented the HS for structural optimization. Geem et al. (2009) reviewed the applications of the HS algorithm in the areas of water resources and environmental system optimization including design of water distribution networks, scheduling of multi-location dams, parameter calibration of environmental models, and determination of ecological reserve location. Karahan et al. (2013) proposed a hybrid HS algorithm for the parameter estimation of the nonlinear Muskingum model. Ambia et al. (2015) applied the HS to optimally design the proportional–integral (PI) controllers of a grid-side voltage converter with two additional loops for smooth transition of islanding and resynchronization operations in a distributed generation (DG) system.

Meta-Heuristic and Evolutionary Algorithms for Engineering Optimization, First Edition. Omid Bozorg-Haddad, Mohammad Solgi, and Hugo A. Loáiciga.

10.2 Inspiration of the Harmony Search (HS)

Music is a widely enjoyed entertainment. The HS is a meta-heuristic algorithm inspired by artificial phenomena found in musical compositions that strive for aesthetic perfection. Musicians test different possible mixtures of musical pitches to achieve a pleasing combination. Such a process of search for a fantastic harmony can be simulated numerically to find the optima of optimization problems.

As an example consider a group of musicians playing a saxophone, double bass, and guitar. Assume that there are a certain number of preferable pitches in each musician's memory: saxophonist {Do, Mi, Sol}, double bassist {Ti, Sol, Re}, and guitarist {La, Fa, Do}. If the saxophonist plays note Sol, the double bassist plays Ti, and the guitarist plays Do, together their notes make a new harmony (Sol, Ti, Do). In other words, musicians improvise a new harmony, which may sound better than the existing worst harmony in their memories, in which case the new harmony is included in their memories and the worst harmony is discarded. This procedure is repeated until an optimal harmony is produced.

Musical improvisation is a process of searching for an optimal or sublime harmony by trying various combinations of pitches following any of the following three rules:

1) Playing any one pitch among stored in the memory
2) Playing a random pitch chosen among those in the possible range of pitches even it is not in the memory
3) Playing a pitch that is close to another pitch already in the memory

According to the first rule, a musician chooses one of the pitches stored in its memory. By the second rule a musician uses a random pitch even it is not in its memory. On the basis of the third rule a musician adopts a close pitch to one present in its memory.

Musicians seek the best state (fantastic harmony) determined by aesthetic feeling, just as the optimization algorithm seeks the best state (global optimum) determined by evaluating the fitness function. Musical aesthetics derive from the set of pitches played by each instrument, just as the fitness function evaluation is performed by the set of values assigned to each decision variable. The harmonic desirability is enhanced practice after practice, just as the solution quality is enhanced iteration by iteration.

According to the HS each harmony is a solution of the optimization problem and pitches that determine the desirability of the harmony represent decision variables. Aesthetic criteria resemble the fitness function of the optimization problem. Creating new solutions in an optimization problem is tantamount in HS to improvising new harmonies during musical creation. A new harmony replaces the worst harmony stored in the musician's memory if it is better than

Table 10.1 The characteristics of the HS.

General algorithm (see Section 2.13)	Harmony search
Decision variable	Pitch
Solution	Harmony
Old solution	Memorized harmony
New solution	New harmony
Best solution	–
Fitness function	Aesthetic criteria
Initial solution	Random harmony
Selection	Updating memory
Process of generating new solutions	Improvising new harmony

the worst harmony. The selection of solutions during the optimization process is analogous to updating the musical memory. Table 10.1 lists the characteristics of the HS.

The HS starts by generating several random harmonies as initial solutions (see Section 2.6) and they are memorized. The fitness values of all the initial solutions are evaluated. These solutions are sorted according to their fitness values, and the worst one is determined. A new harmony is made by musical improvisation. If the new harmony is better than the worst one stored in the memory, the memory is updated and the new harmony replaces the worst one in the memory. Otherwise, the memory is not changed and another new harmony is generated. The process of generating new harmonies and comparing them with the worst memorized harmony is repeated until the termination criteria are satisfied. Figure 10.1 illustrates the flowchart of the HS.

10.3 Initializing the Harmony Memory

Each possible solution of the optimization problem calculated by HS is called a harmony. In other words, harmony's pitches in an N-dimensional optimization problem are the decision variables of the optimization problem. A harmony is known as an array of size $1 \times N$ harmony pitches. This array is defined as follows:

$$Harmony = X = \left(x_1, x_2, \ldots, x_i, \ldots, x_N\right) \tag{10.1}$$

where X = a solution of optimization problem, x_i = ith decision variable of solution X, and N = number of decision variables. The HS algorithm starts with the random generation of a matrix of harmonies of size $M \times N$ (where M and N denote

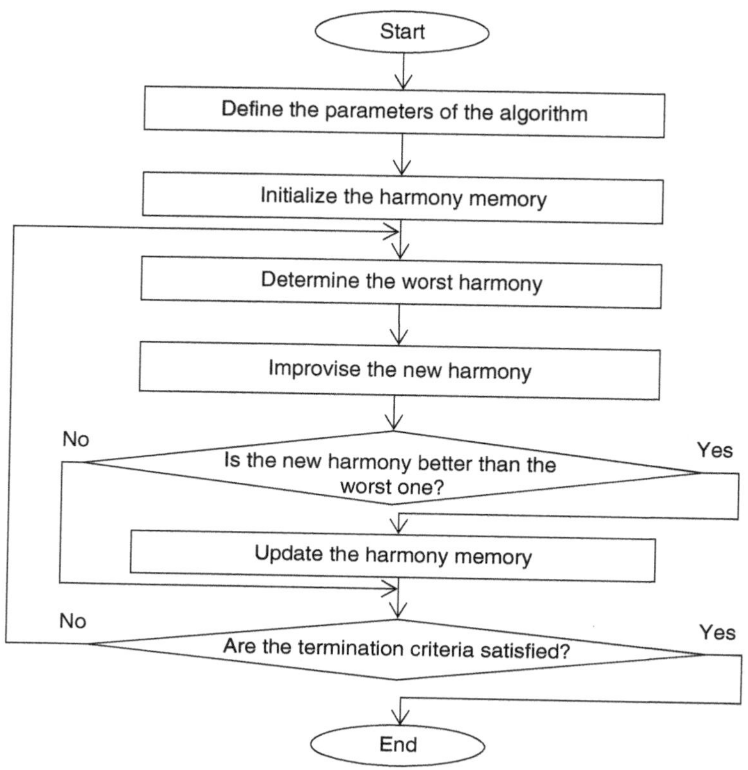

Figure 10.1 The flowchart of the HS.

the size of the harmony memory (HM) and the number of decision variables, respectively). Hence, the matrix of initial possible solutions, which is generated randomly, is as follows (there are M rows or solutions each with N decision variables):

$$Memory = \begin{bmatrix} X_1 \\ X_2 \\ \vdots \\ X_j \\ \vdots \\ X_M \end{bmatrix} = \begin{bmatrix} x_{1,1} & x_{1,2} & \cdots & x_{1,i} & \cdots & x_{1,N} \\ x_{2,1} & x_{2,2} & \cdots & x_{2,i} & \cdots & x_{2,N} \\ & & & \vdots & & \\ x_{j,1} & x_{j,2} & \cdots & x_{j,i} & \cdots & x_{j,N} \\ & & & \vdots & & \\ x_{M,1} & x_{M,2} & \cdots & x_{M,i} & \cdots & x_{M,N} \end{bmatrix} \tag{10.2}$$

in which $Memory$ = HM, X_j = jth solution, $x_{j,i}$ = ith decision variable of the jth solution, and M = capacity of HM. In the HS, each of the decision variable values $(x_1, x_2, x_3, \ldots, x_N)$ is represented as discrete numbers. The HS solves problems with discrete decision space. Each decision variable i takes a value from a predefined set of values V_i as follows:

$$V_i = \left\{ v_{i,1}, v_{i,2}, \ldots, v_{i,d}, \ldots, v_{i,D_i} \right\}, \quad i = 1,2,\ldots,N \tag{10.3}$$

in which V_i = set of possible values for ith decision variable, $v_{i,d}$ = dth possible value for the ith decision variable, and D_i = total number of possible values for the ith decision variable.

10.4 Generating New Harmonies (Solutions)

New solutions have to be produced in the search for optima. The HS search process of generating new solutions is known as improvising harmony, which is performed on the basis of the three previously listed rules of musical improvisation. Each iteration of the HS generates only one new solution even though it is a population-based algorithm.

A value is chosen for each decision variable among the possible values for that decision variable for the purpose of generating a new solution. Choosing values for decision variables is performed by three rules of improvisation in the HS optimization process: (1) memory strategy, (2) random selection, and (3) adjustment pitch. $X^{(new)}$ is a new solution and it is defined as follows:

$$X^{(new)} = \left(x'_1, x'_2, \ldots, x'_i, \ldots, x'_N \right) \tag{10.4}$$

in which $X^{(new)}$ = new solution and x'_i = ith decision variable of the new solution.

10.4.1 Memory Strategy

Memory strategy chooses decision variables' values from those that are stored in the memory. Memory strategy selects randomly one of the memorized solutions for each decision variable and the new solution's value is assigned as follows:

$$x'_i = x_{j,i}, \quad i = 1,2,\ldots,N \tag{10.5}$$

$$j = Irnd(1, M) \tag{10.6}$$

in which $x_{j,i}$ = ith decision variable of the jth solution that is stored in the memory and $Irnd(1, M)$ = an integer random number in the range $[1,M]$. Notice that for each decision variable i, a different random number is generated, and this is done for all the decision variables.

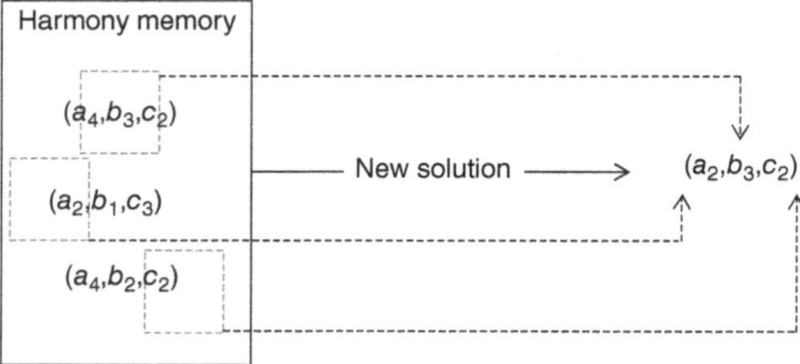

Figure 10.2 Generating a new solution based on memory strategy.

Memory strategy generates a new solution from previously memorized solutions. This strategy for generating new solution is similar to crossover in the genetic algorithm, which combines the decision variables of two previous solutions to form a new solution. However, the memory strategy of the HS is likely to involve more than two solutions in the generation of a new one.

Imagine a three-dimensional problem that includes three decision variables x_1, x_2, and x_3. The possible values for decision variables x_1, x_2, and x_3 are $\{a_1,a_2,a_3,a_4\}$, $\{b_1,b_2,b_3,b_4\}$, and $\{c_1,c_2,c_3,c_4\}$. If the size of HM is equal to 3 ($M=3$) and three solutions are stored in the memory, as depicted in Figure 10.2, solution (a_2,b_3,c_2) would be a new solution on the basis of memory strategy.

10.4.2 Random Selection

Random selection lets decision variables take values that are not in the HM. A new solution is first generated with memory strategy. One or more decision variables are then selected probabilistically to replace their values with random numbers. This approach involves the so-called harmony memory considering rate (HMCR), which is a user-defined parameter of the algorithm and ranges from 0 to 1. The HMCR determines the probability of selecting a decision variable for random replacement. Specifically, for each decision variable a random number in the range [0,1] is generated and compared with the HMCR. The decision variable is selected for random replacement if the randomly generated number is larger than the HMCR. Otherwise, it is not selected. If the ith decision variable is selected for random replacement, its value is determined as follows:

$$x'_i = v_{i,d} \tag{10.7}$$

$$d = Irnd(1, D_i) \tag{10.8}$$

in which $v_{i,d}$ = possible dth value for the ith decision variable.

Consider the example shown in Figure 10.2. If decision variable 3 is selected for random replacement, the new solution may be (a_2, b_3, c_4) in which c_4 is randomly replaces c_2 even though c_4 is not in the memory.

Equation (10.9) determines the decision variables of a new solution considering memory strategy and random selection together.

$$x_i' = \begin{cases} x_{j,i} & \text{if } Rand \le \delta \\ v_{i,d} & \text{if } Rand > \delta \end{cases}, \quad i = 1,2,\ldots,N \tag{10.9}$$

$$j = Irnd(1, M) \tag{10.10}$$

$$d = Irnd(1, D_i) \tag{10.11}$$

in which $Rand$ = a random number in the range [0,1] and δ = HM considering rate (HMCR). For example, an HMCR of 0.95 indicates that the HS algorithm will choose the decision variable value from historically stored values in the memory with a 95% probability or from the entire possible range with a 100 − 95 = 5% probability.

10.4.3 Pitch Adjustment

Pitch adjustment refines newly generated solutions. This is accomplished by examining every component obtained by the memory strategy to determine whether or not it should be pitch adjusted. This operation employs a parameter called the pitch adjustment parameter (PAR) that is user specified and ranges from 0 to 1. Pitch adjustment is a probabilistic process applied to decision variables. A random value is generated in the range [0,1]. If the generated random value is less than the PAR, then the value of a decision variable is changed to a neighboring possible value. For example, a PAR of 0.1 indicates that the algorithm will choose a neighboring value with a 10% probability. If the ith decision variable is randomly selected for pitch adjustment and $v_{i,d}$ is its present value, its new value is determined as follows:

$$x_i' = \begin{cases} v_{i,d+1} & \text{if } Rand > 0.5 \\ v_{i,d-1} & \text{if } Rand \le 0.5 \end{cases} \tag{10.12}$$

in which $v_{i,d+1}$ and $v_{i,d-1}$ = neighboring possible values of $v_{i,d}$ listed as possible values for the ith decision variable. In the example shown in Figure 10.2 a_1 or a_3 can replace a_2 in the new solution if the decision variable 1 is chosen for pitch adjustment.

10.5 Updating the Harmony Memory

The new solution may or may not be selected to enter the HM after it is generated. The new solution replaces the worst one if it has a better fitness value than the worst solution that is stored in the HM. Otherwise, the HM remains unchanged.

10.6 Termination Criteria

The termination criterion determines when to end the algorithm. Selecting a good termination criterion has an important role in the correct convergence of the algorithm. The number of iterations, the amount of improvement of the objective function between consecutive iterations, and the run time are common convergence criteria for the HS.

10.7 User-Defined Parameters of the HS

The size of HM (M), the HM considering rate (HMCR), the PAR, and the criterion used to terminate the algorithm are the user-defined parameters of the HS. A good choice of the parameters is related to the decision space of a particular problem, and usually the optimal parameter setting for one problem is of limited utility for any other problem. Consequently, determining a good parameter setting often requires the execution of many experimental trials. Lots of practice and experience with the HS problems is helpful. A reasonable method for finding appropriate values for the parameters is performing sensitivity analysis, whereby combinations of parameters are tested and the algorithm is run several times for each combination to account for the random nature of the solution algorithm. In this manner the analyst obtains a distribution of solutions and associated objective function values for each combination of parameters. A comparison of the results from all the combination of parameters provides guidance on a proper choice of the algorithmic parameters.

10.8 Pseudocode of the HS

```
Begin
 Input the parameters of the algorithm and initial data
 Generate M initial possible solutions randomly
 Memorize all solutions in the harmony memory
 Evaluate fitness value for all solutions
 While (termination criteria are not satisfied)
    Determine the worst solution in the harmony memory
    For i = 1 to N
        Generate Rand randomly from the range [0,1]
        If Rand > the harmony memory considering
          rate (HMCR)
            Put x'_i = a random value
        Otherwise
            Generate integer number j randomly from
              the range [1,M]
```

```
            Put x'_i = x_{j,i}
            Generate Rand randomly from the range [0,1]
            If Rand ≤ pitch adjustment parameter (PAR)
                Generate Rand randomly from the range
                  [0,1]
                If Rand > 0.5
                    Put x'_i = an upper adjacent value
                Otherwise
                    Put x'_i = an lower adjacent value
                End if
            End if
        End if
    Next i
    Construct new solution X^(new) = (x'_1, x'_2, ..., x'_i, ..., x'_N)
    If the new solution is better than the worst one
      in the harmony memory
        Update the harmony memory
    End if
  End while
  Report the harmony memory
End
```

10.9 Conclusion

This chapter described HS, which is a meta-heuristic algorithm for discrete optimization. First, a brief literature review of the HS was presented. This was followed by a description of the fundamentals of the HS and its algorithmic steps. A pseudocode closed this chapter.

References

Ambia, M. N., Hasanien, H. M., Al-Durra, A., and Muyeen, S. M. (2015). "Harmony search algorithm-based controller parameters optimization for a distributed-generation system." IEEE Transactions on Power Delivery, 30(1), 246–255.

Geem, Z. W., Kim, J. H., and Loganathan, G. V. (2001). "A new heuristic optimization algorithm: Harmony search." Simulation, 76(2), 60–68.

Geem, Z. W., Kim, J. H., and Loganathan, G. V. (2002). "Harmony search optimization: Application to pipe network design." International Journal of Modelling and Simulation, 22(2), 125–133.

Geem, Z. W., Tseng, C. L., and Williams, J. C. (2009). "Harmony search algorithms for water and environmental systems." In: Geem, Z. W. (Ed.), Music-inspired harmony search algorithm, Studies in computational intelligence, Vol. 191, Springer, Berlin, Heidelberg, 113–127.

Karahan, H., Gurarslan, G., and Geem, Z. W. (2013). "Parameter estimation of the nonlinear Muskingum flood-routing model using a hybrid harmony search algorithm." Journal of Hydrologic Engineering, 18(3), 352–360.

Kim, J. H., Geem, Z. W., and Kim, E. S. (2001). "Parameter estimation of the nonlinear Muskingum model using harmony search." Journal of the American Water Resources Association, 37(5), 1131–1138.

Lee, K. S. and Geem, Z. W. (2004). "A new structural optimization method based on the harmony search algorithm." Computers and Structures, 82(9–10), 781–798.

11

Shuffled Frog-Leaping Algorithm

Summary

This chapter describes the shuffled frog-leaping algorithm (SFLA), which is a swarm intelligence algorithm based on the memetic evolution of the social behavior of frogs.

11.1 Introduction

The shuffled frog-leaping algorithm (SFLA) is a swarm intelligence algorithm based on the social behavior of frogs. It was proposed by Eusuff and Lansey (2003). Eusuff et al. (2006) demonstrated the capability of the SFLA for calibrating groundwater models and to design water distribution networks problems. They also compared the results of the SFLA with those of the genetic algorithm (GA). The comparison proved that the SFLA can be an effective tool for solving combinatorial optimization problems. Chung and Lansey (2008) developed a general large-scale water supply model to minimize the total system cost by integrating a mathematical supply system representation applying the SFLA. The results showed that the SFLA found solutions that satisfied all the constraints for the studied networks. Seifollahi-Aghmiuni et al. (2011) implemented the SFLA to analyze the efficiency of a designed network based on nodal demand uncertainty during the operational period. Zhao et al. (2011) presented a combined water quality assessment model constructed based on artificial neural network (ANN) and the SFLA, which was applied to train the initialized data from water quality criteria. Balamurugan (2012) applied the SFLA to achieve the optimum solution of economic dispatch problem with multiple fuel options and demonstrated that the SFLA algorithm provides quality solutions with less computational time than other techniques reported in the literature. Fallah-Mehdipour et al. (2013) extracted multi-crop planning rules in a reservoir

Meta-Heuristic and Evolutionary Algorithms for Engineering Optimization, First Edition. Omid Bozorg-Haddad, Mohammad Solgi, and Hugo A. Loáiciga.

system with the SFLA algorithm, the GA, and the particle swarm optimization (PSO). Orouji et al. (2013) compared the performance of the SFLA with the simulated annealing (SA) in estimation of Muskingum flood routing parameters. The result showed the superiority of the SFLA relative to the SA. Seifollahi-Aghmiuni et al. (2013) applied the SFLA to evaluate performance of a water distribution network under pipe roughness uncertainty during an operational period. Orouji et al. (2014) proposes a hybrid algorithm, based on the PSO and SFLA, to solve the resource-constrained project scheduling problem (RCPSP), which aims at the minimization of time required to complete a project considering resource limitations and the timing of activities. Results showed that the hybrid PSO–SFLA is quite capable to determine an optimal solution in all problems, even with a fewer number of iterations compared with the individual application of the PSO and SFLA. Bozorg-Haddad et al. (2015) proposed a novel hybrid algorithm, based on the SFLA and the Nelder–Mead simplex (NMS), for the estimation of parameters of two new nonlinear Muskingum flood routing models. Mahmoudi et al. (2016) presented a novel tool for estimation of quality of surface water by coupling support vector regression (SVR) and the SFLA. Their results indicated that the new proposed SFLA–SVR tool is more efficient and powerful tool for determining water quality parameters in comparison with other previously methods such as genetic programming (GP).

11.2 Mapping Memetic Evolution of Frogs to the Shuffled Frog Leaping Algorithm (SFLA)

A meme is a spreading information template that affects human and animal minds and changes their behavior. Memes are spread by those who possess them. A pattern is known as a meme whenever an idea or information template influences someone, and the template is repeated or transmitted to someone else. Otherwise, the pattern is not a meme. Notice that all transmitted information is called memetic. Examples of memes are songs, ideas, catch phrases, clothing fashions, and techniques for making pots or building arches.

A memetic algorithm (MA), which derives from "meme," is a population-based method to solve optimization problems (Eusuff et al., 2006). Each meme contains memotypes that resemble the genes of a chromosome. Memes spread through the meme pool as they move from brain to brain. Genes and memes scatter from one individual to another in various ways, and their purposes are different. Memes are used basically for increased communicability among their hosts (described as frogs in the SFLA). Genes transmit DNA characteristics from parents to offspring.

Eusuff et al. (2006) stated that memetic and genetic evolution are subjected to the same basic principles. Yet, memetic evolution is a much more flexible mechanism than genetic evolution. They reasoned that genes can only be transferred from parents to offspring and are transmitted between generations, meaning that their propagation through higher organisms may take several years to propagate. Memes, on the other hand, can theoretically be transformed between any two individuals and can be transmitted within minutes. Gene replication is restricted by the relatively small number of offspring from a single parent, whereas the number of individuals that can take over a meme from a single individual is almost unlimited. Therefore, meme spreading is much faster than gene spreading (Eusuff and Lansey, 2003).

Think of a group of frogs leaping in a swamp. There are stones at different locations throughout the swamp. Frogs want to find the stone with the maximum amount of available food as fast as possible. For this purpose they improve their memes. The frogs interact with each other and develop their memes by exchanging information. Frogs change their positions by adjusting their leaping step size based on the development of memes.

The SFLA acts as an MA that progresses by transforming frogs in memetic evolution. Individual frogs of the SFLA are hosts for memes and are represented by means of a memetic vector. Made of a number of memotypes, each meme attributed to a frog is a solution of an optimization problem, while memotypes are the decision variables and resemble the genes of a chromosome in the genetic algorithm. The SFLA does not change the physical characteristics of an individual; rather, it progressively improves the ideas held by each frog in a so-called virtual population, which is used to model the meme pool consisting of a diverse set of frogs in a manner analogous to the population representing a chromosome pool in a GA population. A set of frogs represents a population of solutions. The population of possible solutions is partitioned into subsets that are called memeplexes. Each meme is the unit of cultural evolution. The term memeplex is introduced to mark a group of mutually supporting memes that form an organized belief system, such as a religion or scientific theory. The memeplexes can be perceived as a set of parallel frog cultures attempting to reach some goal. Each frog culture or memeplex evolves toward its goal. Frog leaping improves an individual's meme and enhances its performance toward the goal. Within each memeplex the frogs are influenced by other frogs' ideas. Hence they experience a memetic evolution. Information is passed between memeplexes in a shuffling process according to memetic evolutionary steps. Table 11.1 lists the characteristics of the SFLA.

The SFLA starts the optimization process by randomly generating a set of frog memes (see Section 2.6), each of which is a solution of the optimization problem. All the initially generated frogs (or solutions) are classified into several memeplexes, so that each frog is assigned to one memeplex. These memeplexes allow evolving independently by searching the solution space in different directions. Information is then passed between memeplexes in a shuffling process. The search for the optimal solutions by the memeplexes continues after shuffling. The searches by the memeplexes and the shuffling process continue until the defined termination criteria are satisfied. Figure 11.1 depicts the flowchart of the SFLA.

Table 11.1 The characteristics of the SFLA.

General algorithm (see Section 2.13)	Shuffled frog leaping algorithm
Decision variable	Memotype
Solution	Meme of frog
Old solution	Previous meme of frog
New solution	Improved meme of frog
Best solution	Best frog
Fitness function	Amount of food
Initial solution	Random frog
Selection	Classification frogs into memeplexes
Process of generating new solution	Frog leaping

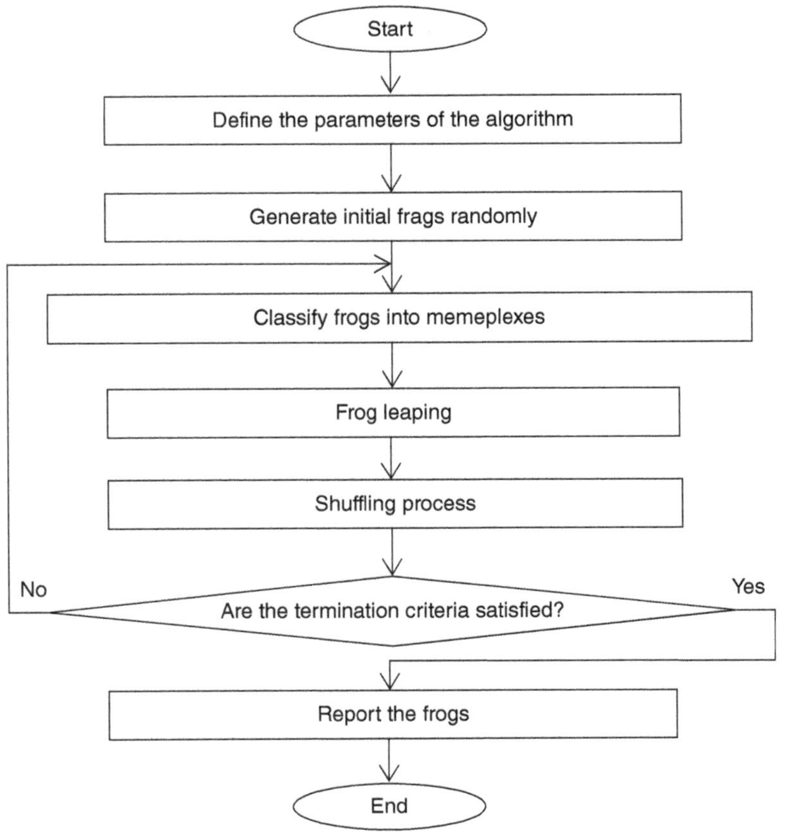

Figure 11.1 The flowchart of the SFLA.

11.3 Creating an Initial Population

Each possible solution of the optimization problem is called a frog in the SFLA. Each solution or frog contains N decision variables or memotypes when solving an N-dimensional optimization problem. A solution or frog is represented as an array of size $1 \times N$. This array is as follows:

$$Frog = X = (x_1, x_2, \ldots, x_i, \ldots, x_N) \tag{11.1}$$

where X = a possible solution (frog) of the optimization problem, x_i = ith decision variable (memotype) of solution X, and N = the number of decision variables. The decision variable values $(x_1, x_2, x_3, \ldots, x_N)$ designate the memotypes.

The SFLA begins with the random generation of a matrix of size $M \times N$ (see Section 2.6) where M and N denote the size of the population of solutions and the number of decision variables, respectively. The matrix of solutions generated randomly is represented as follows (each row contains the decision variables of a solution or frog, and there are M rows):

$$Population = \begin{bmatrix} X_1 \\ X_2 \\ \vdots \\ X_j \\ \vdots \\ X_M \end{bmatrix} = \begin{bmatrix} x_{1,1} & x_{1,2} & \cdots & x_{1,i} & \cdots & x_{1,N} \\ x_{2,1} & x_{2,2} & \cdots & x_{2,i} & \cdots & x_{2,N} \\ & & & \vdots & & \\ x_{j,1} & x_{j,2} & \cdots & x_{j,i} & \cdots & x_{j,N} \\ & & & \vdots & & \\ x_{M,1} & x_{M,2} & \cdots & x_{M,i} & \cdots & x_{M,N} \end{bmatrix} \tag{11.2}$$

in which X_j = jth solution, $x_{j,i}$ = ith decision variable of the jth solution, and M = population size.

11.4 Classifying Frogs into Memeplexes

A fitness function is used to evaluate the worth of each frog. Frogs are then sorted based on the values of their fitness function in ascending or descending order in a minimization or maximization problem, respectively. Figure 11.2 depicts the sorting of frogs according to the values of the fitness function $F(X)$ in a maximizing problem.

After sorting the frogs are assigned to Z memeplexes, with Y frogs in each memeplex. The frog with the best value of the fitness function becomes the first member of the first memeplex; the second-best frog becomes the first member of the second memeplex. The assignment of sorted frogs continues until the Zth-sorted frog becomes the first member of the Zth memeplex. In the next step, the $(Z+1)$st frog becomes the second member of the first memeplex and so on. Figure 11.3 portrays the assignment of frogs to the memeplexes. Y and Z

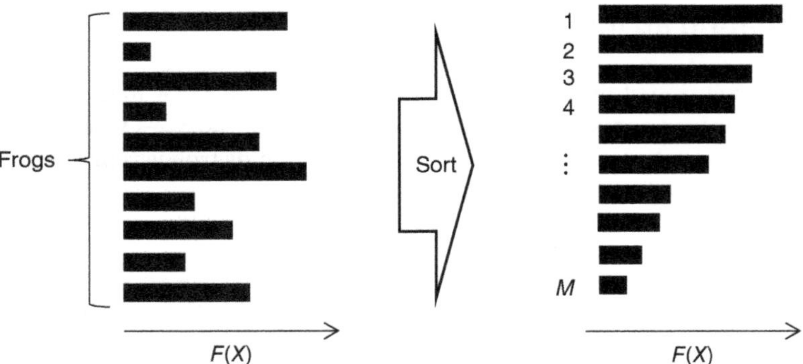

Figure 11.2 Sorting frogs according to the fitness function $F(X)$ in a maximizing problem.

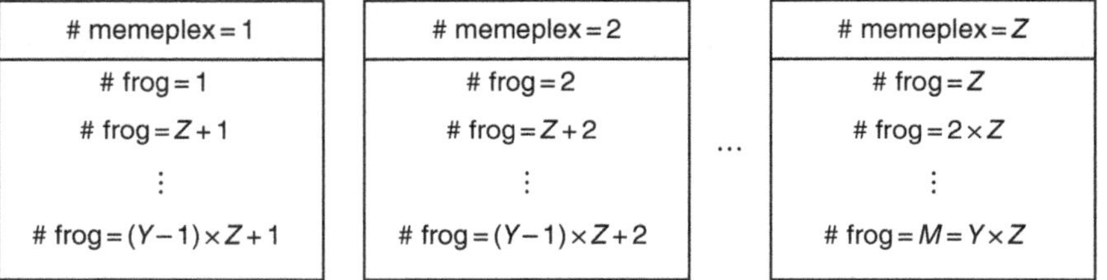

Figure 11.3 Assigning frogs to different memeplexes; Z = number of memeplexes and Y = number of frogs assigned to each memeplex.

are user-defined parameters of the algorithm, and their values are determined by user. Consequently, the population size (M) is equal to $Z \times Y$.

11.5 Frog Leaping

After the population of possible solutions of frogs is classified into several parallel communities (memeplexes), the frogs are influenced by other frogs' ideas within each memeplex generating a so-called memetic evolution. Memetic evolution improves the quality of the meme of an individual and enhances the individual frog's performance toward a goal. The frogs with better memes (ideas) contribute more to the development of new ideas than frogs with poor ideas. This ensures that memetic evolution (i.e., the spreading of superior ideas) selects with higher probability the best individuals to continue the search for optima.

The frogs' purpose is to move toward the optimum by improving their memes. For this purpose for each memeplex, a subset of the memeplex called a submemeplex is chosen for the transmission of ideas. In actuality, the submemeplex has $Q < Y$ frogs. The concept of a submemeplex is depicted in Figure 11.4.

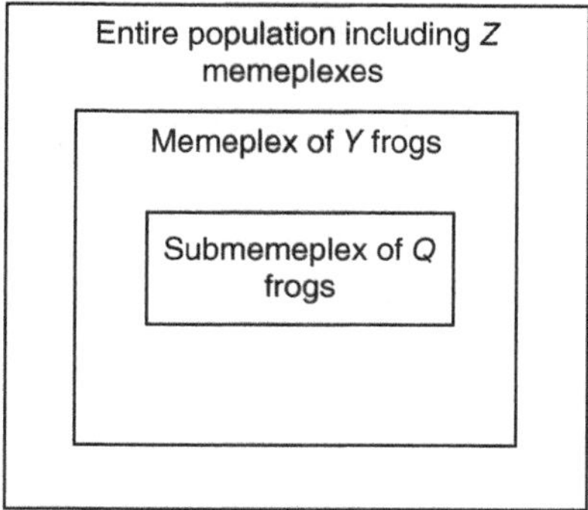

Figure 11.4 The representation of a memeplex and a submemeplex within the entire population.

The choice of frogs to form a submemeplex conforms to the strategy that superiorly fit frogs have a higher probability of forming the submemeplex. The frogs of a memplex are ranked according to their fitness functions. The probability with which the Q frogs are selected from a memeplex to form a submemeplex conforms to the following distribution:

$$P_j = \frac{2 \times (Y + 1 - j)}{Y \times (Y + 1)}, \quad j = 1,2,\ldots,Y \tag{11.3}$$

in which P_j = the probability that the jth frog of a memeplex is selected for the submemeplex, j = counter of frogs in the memeplex, and Y = total number of frogs in the memeplex. $j = 1$ denotes the best frog in the memplex, and $j = Y$ denotes the worst frog in the memeplex. It follows from Equation (11.3) that the fittest ($j = 1$) frog of the memeplex has a probability equal to $2/(Y+1)$ of being selected, and the least fit frog ($j = Y$) of the memeplex is chosen with a probability $2/(Y^2 + Y)$.

Q frogs are selected to form a submemeplex, and they are ranked with the fittest frog having rank 1 and the worst frog having rank Q according to their fitness values. Next, the worst frog in the submemeplex is improved. The worst frog adopts its meme from the best frog within the submemeplex and from the globally best frog. The best and worse frogs in each submemeplex and the best global frog in the entire population of frogs are called *Mbest*, *Mworst*, and *Pbest*, respectively. The meme of the *Mworst* is improved as follows:

$$d_i = Rand \times \left(x_{Mbest,i} - x_{Mworst,i}\right), \quad D_{min} \le d_i \le D_{max}, \quad i = 1,2,\ldots,N \tag{11.4}$$

$$x'_i = x_{Mworst,i} + d_i, \quad i = 1,2,\ldots,N \tag{11.5}$$

$$X^{(new)} = \left(x'_1, x'_2, \ldots, x'_i, \ldots, x'_N\right) \tag{11.6}$$

where d_i = leaping step size for the ith decision variable of the worst frog in the submemeplex; $Rand$ = random values in the range [0,1]; $x_{Mbest,i}$ = ith decision variable of the best solution in the submemeplex; $x_{Mworst,i}$ = ith decision variable of the worst solution in the submemeplex; D_{min} and D_{max} = minimum and maximum allowable values for the leaping step size, respectively; x'_i = ith decision variable of the new solution; and $X^{(new)}$ = the new solution.

If $X^{(new)}$ is better than the *Mworst*, it replaces *Mworst*. Otherwise, $X^{(new)}$ is generated based on *Pbest* instead of *Mbest* as follows:

$$d_i = Rand \times \left(x_{Pbest,i} - x_{Mworst,i}\right),\quad D_{min} \le d_i \le D_{max},\quad i = 1,2,\ldots,N \tag{11.7}$$

$$x'_i = x_{Mworst,i} + d_i,\quad i = 1,2,\ldots,N \tag{11.8}$$

$$X^{(new)} = \left(x'_1, x'_2, \ldots, x'_i, \ldots, x'_N\right) \tag{11.9}$$

in which $x_{Pbest,i}$ = ith decision variable of the best solution in the entire population.

In case $X^{(new)}$ is not better than *Mworst*, a randomly generated possible solution substitutes *Mworst* as follows:

$$x'_i = Rnd\left(x_i^{(L)}, x_i^{(U)}\right),\quad i = 1,2,\ldots,N \tag{11.10}$$

$$X^{(new)} = \left(x'_1, x'_2, \ldots, x'_i, \ldots, x'_N\right) \tag{11.11}$$

in which $Rnd(a,b)$ = a random value from the range $[a,b]$, $x_i^{(L)}$ and $x_i^{(U)}$ = the lower and upper allowable values of the ith decision variable, respectively.

The frogs of the memeplex are ranked according to their fitness values following the replacement of *Mworst*. A new set of solutions are selected randomly to construct a new submemeplex, and the worst solution among them is updated as described previously. This process is executed for all the memeplexes for a user-defined number of times μ. After a certain number of memetic evolutions (μ), the memeplexes are forced to mix to form new memeplexes through a shuffling process. This shuffling enhances the quality of the memes after being influenced by the ideas of frogs from different regions of the swamp (different memeplexes).

11.6 Shuffling Process

The intercultural migration of frogs accelerates the searching procedure through the sharing of information, and it ensures that the cultural evolution toward any particular goal or optimum is free from regional or group bias. In fact, shuffling guarantees that there is no bias in cultural evolution toward any specific goal (Eusuff et al., 2006).

Shuffling leads to the convergence of cultures that evolve in isolation until they are brought together to share ideas or information. Shuffling mixes all the memeplexes into a unique population of possible solutions from which next memeplexes are produced. Also, the shuffling process updates the best frog.

11.7 Termination Criteria

The termination criterion determines when to end the SFLA. Selecting a good termination criterion has an important role on the correct convergence of the algorithm. The number of iterations, the amount of fitness improvement of the solution between consecutive iterations, the number of algorithmic iterations, and the run time are common termination criteria for the SFLA.

11.8 User-Defined Parameters of the SFLA

The number of memeplexes (Z), the number of solutions in each memeplexes (Y), the submemeplex size (Q), the minimum and maximum allowable values for the leaping step size, the number of memetic evolutions (μ), and the criterion used to decide when to terminate the algorithm are user-defined parameters of the SFLA. A good choice of the parameters depends on the decision space of a particular problem, and frequently the optimal parameter setting for one problem is of limited utility for any other problem. Consequently, determining a good parameter setting often requires experimentation with the parameter set and experience with specific problems. A reasonable method for finding appropriate values for the parameters is performing sensitivity analysis. This entails choosing various sets of parameters and running the SFLA with each set of parameters a number of times to account for the random nature of its solution scheme. This is repeated for each set of parameters, and carrying out a comparison of the solutions from all runs to gain an insight on an appropriate choice of algorithmic parameters.

11.9 Pseudocode of the SFLA

```
Begin
  Input the parameters of the algorithm and initial data
  Let Z = the number of memeplexes; Y = the number of
    frogs in each memeplex; μ = the number of memetic
    evolutions; Q = submemeplex size; and M = Y × Z
  Generate M initial possible frogs (solutions) randomly
    and evaluate their fitness values
  While (the termination criteria are not satisfied)
     Sort the frogs according to their fitness values
     Divide frogs into Z memeplexes each of which has
       Y frogs
     For s = 1 to Z
         For j = 1 to μ
```

```
                Choose Q frogs from memeplex s randomly
                  and make a submemeplex
                Improve the worst frog of the submemeplex
                  (MWorst) according to the best frog of
                  the submemeplex (MBest) and update its
                  fitness value
                If there is no improvement in the worst frog
                      Improve the worst frog of the
                        submemeplex (MWorst) according to
                        the best frog in the population
                        (PBest) and update its fitness
                        value
                End if
                If there is no improvement in the worst frog
                      Generate a random frog to replace the
                        worst frog of the submemeplex and
                        evaluate its fitness value
                End if
            Next j
        Next s
        Combine all memeplexes.
    End while
    Report the best solution
End
```

11.10 Conclusion

This chapter described the SFLA, which is a swarm intelligence algorithm based on the memetic evolution of the social behavior of frogs. The chapter presented a brief literature review of the SFLA and its applications, its algorithmic fundamentals, and a pseudocode.

References

Balamurugan, R. (2012). "Application of shuffled frog leaping algorithm for economic dispatch with multiple fuel options." International Conference on Emerging Trends in Electrical Engineering and Energy Management (ICETEEEM), Chennai, Tamil Nadu, India, December 13–15, Piscataway, NJ: Institute of Electrical and Electronics Engineers (IEEE).

Bozorg-Haddad, O., Hamedi, F., Fallah-Mehdipour, E., Orouji, H., and Mariño, M. A. (2015). "Application of a hybrid optimization method in Muskingum parameter estimation." Journal of Irrigation and Drainage Engineering, 141(12), 04015026.

Chung, G. and Lansey, K. (2008). "Application of the shuffled frog leaping algorithm for the optimization of a general large-scale water supply system." Water Resources Management, 23(4), 797–823.

Eusuff, M. M. and Lansey, K. E. (2003). "Optimization of water distribution network design using the shuffled frog leaping algorithm." Journal of Water Resources Planning and Management, 129(3), 210–225.

Eusuff, M. M., Lansey, K. E., and Pasha, F. (2006). "Shuffled frog-leaping algorithm: A memetic meta-heuristic for discrete optimization." Engineering Optimization, 38(2), 129–154.

Fallah-Mehdipour, E., Bozorg-Haddad, O., and Mariño, M. A. (2013). "Extraction of multi-crop planning rules in a reservoir system: Application of evolutionary algorithms." Journal of Irrigation and Drainage Engineering, 139(6), 490–498.

Mahmoudi, N., Orouji, H., and Fallah-Mehdipour, E. (2016). "Integration of shuffled frog leaping algorithm and support vector regression for prediction of water quality parameters." Water Resources Management, 30(7), 2195–2211.

Orouji, H., Bozorg-Haddad, O., Fallah-Mehdipour, E., and Mariño, M. A. (2013). "Estimation of Muskingum parameterby meta-heuristic algorithms." Proceedings of the Institution of Civil Engineers, Water Management, 165(1), 1–10.

Orouji, H., Bozorg-Haddad, O., Fallah-Mehdipour, E., and Mariño, M. A. (2014). "Extraction of decision alternatives in project management: Application of hybrid PSO-SFLA." Journal of Management in Engineering, 30(1), 50–59.

Seifollahi-Aghmiuni, S., Bozorg-Haddad, O., Omid, M. H., and Mariño, M. A. (2011). "Long-term efficiency of water networks with demand uncertainty." Proceedings of the Institution of Civil Engineers: Water Management, 164 (3), 147–159.

Seifollahi-Aghmiuni, S., Bozorg-Haddad, O., Omid, M. H., and Mariño, M. A. (2013). "Effects of pipe roughness uncertainty on water distribution network performance during its operational period." Water Resources Management, 27(5), 1581–1599.

Zhao, Y., Dong, Z. C., and Li, Q. H. (2011). "ANN based on SFLA for surface water quality evaluation model and its application." 2011 International Conference on Transportation, Mechanical, and Electrical Engineering (TMEE), Changchun, China, December 16–18, Piscataway, NJ: Institute of Electrical and Electronics Engineers (IEEE), 1615–1618.

12

Honey-Bee Mating Optimization

Summary

This chapter describes the honey-bee mating optimization (HBMO) algorithm, which is based on the mating strategy of honey bees. The chapter presents a review of the HBMO, its applications, fundamentals, algorithmic steps, and a pseudocode.

12.1 Introduction

Honey bees are social insects that live in large and well-organized hives. Social intelligence, observance of collective rules, and division of labor are some of the traits that honey bees exhibit. Honey bees mate and reproduce in a unique way. The honey-bee mating optimization (HBMO) algorithm is inspired by the honey-bee mating process. It was developed and applied to reservoir operation by Bozorg-Haddad et al. (2006). Bozorg-Haddad and Mariño (2007) proposed dynamic penalty function as a strategy in solving water resources combinatorial optimization problems with the HBMO algorithm. Bozorg-Haddad et al. (2009) applied the HBMO to solve non-convex optimization problems. Several studies have reported the successful application of the HBMO algorithm to solve a variety of problems such as water reservoir operation (Afshar et al., 2007; Bozorg-Haddad and Mariño, 2008; Bozorg-Haddad et al., 2008b, 2010a, b; Afshar et al., 2011), water distribution networks (Bozorg-Haddad et al., 2008a; Jahanshahi and Bozorg-Haddad, 2008; Ghajarnia et al., 2009, 2011; Soltanjalili et al., 2011; Sabbaghpour et al., 2012; Soltanjalili et al., 2013a, b; Solgi et al., 2015; Bozorg-Haddad et al., 2016a, b, c; Solgi et al., 2016b), project management (Bozorg-Haddad et al., 2010c), supply chain management (Marinakis et al., 2008a), clustering analysis (Marinakis et al., 2008b), electric distribution systems (Niknam, 2009), image processing and pattern

Meta-Heuristic and Evolutionary Algorithms for Engineering Optimization,
First Edition. Omid Bozorg-Haddad, Mohammad Solgi, and Hugo A. Loáiciga.

recognition (Horng et al., 2009), design and operation of run-of-river power plants (Bozorg-Haddad et al., 2011), and groundwater management (Bozorg-Haddad and Mariño, 2011). Several of those works have proven the superiority of the HBMO algorithm compared with other algorithms such as the genetic algorithm (GA), ant colony optimization (ACO), and particle swarm optimization (PSO) for the chosen applications. Karimi et al. (2013) proved the better performance of the HBMO algorithm than that of the GA solving various test functions. Solgi et al. (2016a) modified the HBMO leading to the enhanced HBMO (EHBMO) algorithm and demonstrated the superiority of the EHBMO on the HBMO and elitist GA in solving several mathematical functions and water resources optimization problems.

12.2 Mapping Honey-Bee Mating Optimization (HBMO) to the Honey-Bee Colony Structure

There is fossil evidence of honey bees' existence dating back 100 million years ago (Michener and Grimaldi, 1988). Honey bees live together in well-organized hives. The purpose of a hive is to maximize the efficiency of the bees by means of the division of the labor. A well-organized hive remains viable except in special circumstances. A colony of bees is a group of bees living together in one bee hive. A honey-bee colony typically consists of a single egg-laying long-lived queen, anywhere from zero to several thousand drones (depending on the season) and usually 10 000–60 000 workers. So a honey-bee hive consists of a single queen, broods, drones, and workers (Moritz and Southwick, 1992). The queen and workers are female, while drones are male. The queen is generally the only bee that can mate with drones and can fertilize the eggs. However, queens are not the only colony members capable of reproduction. Honey-bee workers cannot mate but can lay male eggs. Mate production by workers in the honey bee is rare, however, due to workers' policing. The primary duty of workers is brood caring. Drones are the fathers of the colony. The queen can lay both fertilized and unfertilized eggs. Fertilized eggs represent female bees (worker or queen) and unfertilized eggs represent drones. Drones are haploid and amplify their mother's genome without alteration of their genetic composition except through mutation. However female bees inherit their genome from both their mothers and fathers. When a new queen is born, it replaces the old queen or it leaves the hive.

The queen is the most important member of the hive because she is the one that breeds new members. With the help of approximately 18 males (drones), the queen bee mates from one to five times over several days in her life. The sperm from the drone is planted inside a pouch in her body. She uses the stored sperm to fertilize the eggs. The queen exits from the hive and engages in a mating flight around the hive to fertilize her eggs. In each mating flight the queen usually mates with 7–20 drones. In each mating the drone's sperm reaches the queen's spermatheca and accumulates there to form the genetic pool of the colony. After the end of the mating flight, the queen returns to the hive and starts laying eggs. The successful drones in mating flights die immediately after mating with the queen. In the other words, insemination ends with the death of the drone. The unsuccessful drones (those that do not mate with the queen) also die from starvation and exposure because the workers forbid their entry to the hive at the end of the mating season. Usually, as the nights turn colder and winter arrives, the drones still in the hive are forced out of the hive by worker bees. This is a survivalist sacrifice because the hive would not have enough food if the drones remain in the hive. The queen usually starts laying eggs in the middle of February and continues to do so till the end of June. The population of the hive grows day by day as a result of reproduction until space shortages appear in the hive.

The HBMO algorithm mimics the queen, broods, and drones as possible solutions that are made up of genes. Each gene is equivalent to a decision variable. The best solution is considered as the queen. Broods can be diploid or haploid broods. The former are made by applying mutation and crossover operators on the queen's genome and drone's, whereas the latter are made by applying mutation on the queen. Brood caring by workers is mapped into the algorithm to improve the broods by applying heuristic functions. The queens play the most important role in the mating process in nature as well as in the HBMO algorithm. Each queen is characterized with a genotype, speed, energy, and a spermatheca with defined capacity. Spermatheca is the repository of drones' sperm produced during mating with the queen. Therefore, for a queen with defined spermatheca size, speed and energy are initialized before each mating flight. After successful mating, the drones' sperm is stored in the queens' spermatheca. Later in the breeding process, a brood is constructed by copying some of the drones' genes into the brood genotype and completing the rest of the genes from the queens' genome. The fitness of the resulted genotype is determined by evaluating the value of the fitness function of the brood genotype and/or its normalized value. It is important to note that a brood has only one genotype. A mating flight is mapped into the HBMO algorithm as the queen randomly chooses drones from the decision space of the problem. The genome of each drone that is successful in mating is stored in the queen's spermatheca. Also, the death of drones at the end of the mating season is simulated by destroying all remaining drones after the mating flight in each iteration of the HBMO algorithm. The characteristics of the HBMO algorithm are listed in Table 12.1.

The HBMO algorithm starts with the random generation of the initial population of possible solutions (see Section 2.6). The solutions are ranked based on their fitness values. The fittest (best) solution is marked out as the queen. In the next step a mating flight is implemented to randomly select drones (solutions) from the decision space for mating with the queen. The genome of

Table 12.1 The characteristics of the HBMO.

General algorithm (see Section 2.13)	Honey-bee mating optimization
Decision variable	Gene
Solution	Bee (drone/queen/brood)
Old solution	Queen and drone
New solution	Brood
Best solution	Queen
Fitness function	Fitness of bee
Initial solution	Random bee
Selection	Mating process
Process of generating new solutions	Genetic operators and brood caring

each selected drone is stored in the queen's spermatheca. The remaining solutions are deleted after the mating flight. The queen and the solution stored in the queen's spermatheca are used to make the next generation. First, the broods (diploid or haploid) are made. The haploid broods are made by applying mutation on the queen. The diploid broods are made by applying crossover and mutation operators between the queen and the solutions stored in the queen's spermatheca. Heuristic functions that model worker bees are applied in an attempt to improve the broods. Finally, the best brood replaces the old queen if it is better than the old queen. Figure 12.1 depicts the flowchart of the HBMO algorithm.

12.3 Creating an Initial Population

Each possible solution of the optimization problem calculated by the HBMO algorithm is called a bee. Each bee (drone, queen, or brood) in the mathematical formulation of an optimization problem symbolizes a series of genes (decision variable) that represent a solution of the problem. In an N-dimensional optimization problem, a bee is an array of size $1 \times N$. This array is as follows:

$$Bee = X = \left(x_1, x_2, \ldots, x_i, \ldots, x_N\right) \tag{12.1}$$

where X = a solution of optimization problem, x_i = ith decision variable of solution X, and N = number of decision variables. Each of the decision variable values $(x_1, x_2, x_3, \ldots, x_N)$ can be represented as floating point number (real values) or as a predefined set for continuous and discrete problems, respectively.

A matrix of size $M \times N$ is generated randomly (see Section 2.6), where M and N are the size of the population of solutions and the number of decision

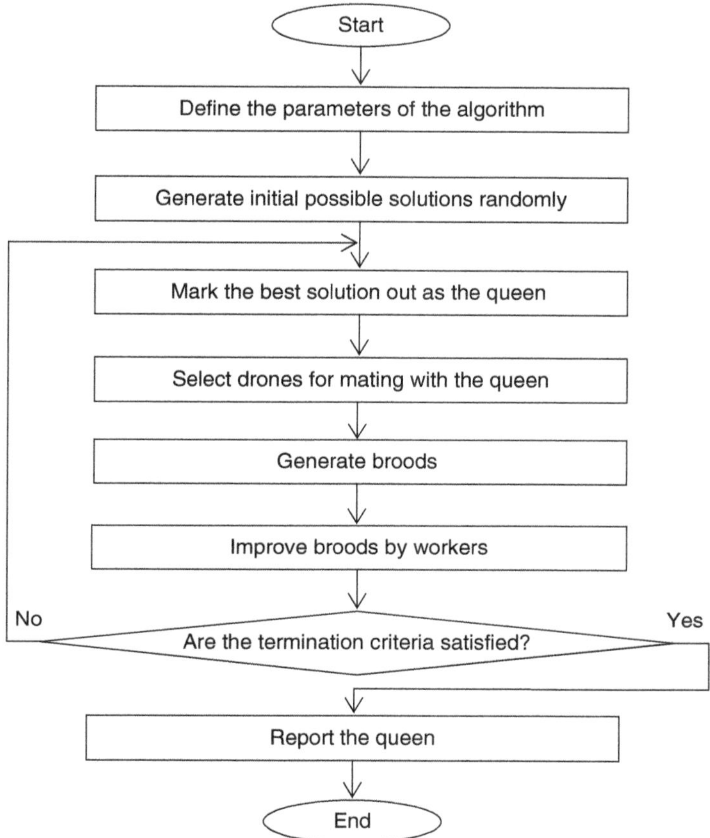

Figure 12.1 The flowchart of the HBMO algorithm.

variables, respectively. Hence, the matrix of solutions that is generated randomly is given as follows (rows and columns are the number of bees and the number of decision variables, respectively):

$$Population = \begin{bmatrix} X_1 \\ X_2 \\ \vdots \\ X_j \\ \vdots \\ X_M \end{bmatrix} = \begin{bmatrix} x_{1,1} & x_{1,2} & \cdots & x_{1,i} & \cdots & x_{1,N} \\ x_{2,1} & x_{2,2} & \cdots & x_{2,i} & \cdots & x_{2,N} \\ & & & \vdots & & \\ x_{j,1} & x_{j,2} & \cdots & x_{j,i} & \cdots & x_{j,N} \\ & & & \vdots & & \\ x_{M,1} & x_{M,2} & \cdots & x_{M,i} & \cdots & x_{M,N} \end{bmatrix} \tag{12.2}$$

in which X_j = jth solution, $x_{j,i}$ = ith decision variable of the jth solution, and M = population size.

12.4 The Queen

As mentioned previously, the queen plays the most important role in the mating process in nature as well as in the HBMO algorithm. Each queen is characterized with a genotype, speed, energy, and a spermatheca with defined capacity. Genotype characterizes the queen as a solution of the optimization problem (the best solution in the present generation). The speed and energy of the queen are parameters of the algorithm that control the rate of convergence and are described in the following sections in detail. The spermatheca is the repository of drones' sperm produced during mating with the queen. The speed and energy of a queen with spermatheca size (S_c) are initialized before each mating flight at random in the range of (0.5,1). All the bees are ranked based on their fitness values. The best solution (i.e., the bee with the best fitness value) is made queen. Figure 12.2 shows how to select the queen in a minimization problem.

It is seen in Figure 12.2 that after all the solutions are ranked, the best one is designated as the queen and other solutions become trial solutions. The rules for trial solutions are described in the following sections.

12.5 Drone Selection

The HBMO selects drones to mate with the queen and generate broods (new solutions). The queen is the mother of all the new solutions, and they have different fathers that are drones selected for mating. Two strategies are used for drone selection: (1) mating flights and (2) considering trial solutions as drones.

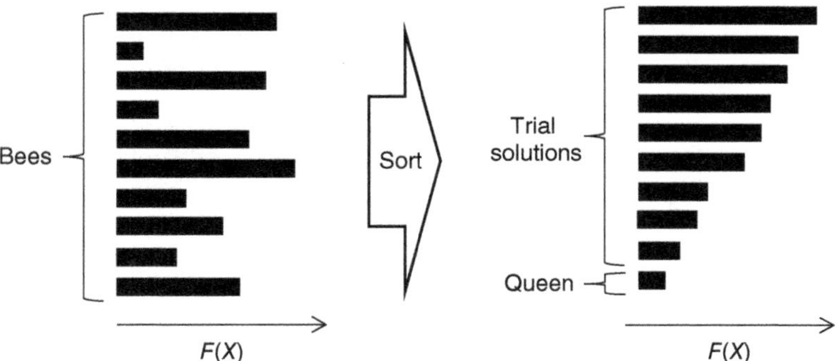

Figure 12.2 Determining the queen and trial solutions according to the fitness function $F(X)$.

12.5.1 Mating Flights

During each mating flight, the queen flies based on its energy and speed, which are generated at random for the queen before each mating flight commences. The mating flight begins, and the queen, based on its speed and energy, finds the best randomly generated drones for mating. The queen's motion is governed by its speed, which is the probability of mating between the queen and drones. The speed of the queen is at its maximum value at the beginning of the mating flight, and it decreases as the queen flies and vets different drones. Therefore, the space of the queen's activity decreases as it continues to fly. This means that the probability of the mating for a drone in the beginning of the mating flight is higher than the probability of mating for the same drone at the end of the mating flight. At every step of the mating flight, the queen tests its adjacent drone using a probability function. The drones' sperm is saved in the queen's sperm bag if the mating is successful (the drone passes the threshold).

Drones are randomly generated for the mating process. A drone's genome is memorized in the queen's spermatheca after evaluating its fitness value if the drone succeeds in a simulated annealing (SA) function as follows:

$$\zeta\left(X_q,X_d\right)=e^{\frac{-\left|F(X_q)-F(X_d)\right|}{\lambda}} \tag{12.3}$$

in which $\zeta(X_q,X_d)$ = the probability of mating a drone with the queen, $F(X_q)$ = the fitness function of the queen, $F(X_d)$ = the fitness function of the drone, and λ = queen's speed, which is a parameter of the algorithm and whose value regulates selective pressure. The selective pressure is high (low) when λ is low (high). A uniformly distributed random variable (*Rand*) within [0,1] is generated while $\zeta(X_q,X_d)$ is evaluated. If $\zeta(X_q,X_d)$ is larger than or equal to *Rand*, the drone is successful in mating with the queen; otherwise it is not. Equation (12.3) acts as an annealing function. Whether or not the mating between a drone and the queen is successful, another drone is randomly generated until the queen's spermatheca (S_c) is full or the queen's energy is finished.

The probability of mating is high when either the queen's speed is high or the fitness of the drone ($F(X_d)$) is as good as that of the queen ($F(X_q)$). It is also high at the beginning of the mating flight when the queen's speed is high or when the drone is fit enough. The queen's energy and speed decrease after each movement of the queen in space or after each mating according to the following equations:

$$\psi^{(new)}=\psi-\gamma \tag{12.4}$$

$$\lambda^{(new)}=\alpha\times\lambda \tag{12.5}$$

where $\psi^{(new)}$ = new energy of the queen, ψ = old energy of the queen, $\lambda^{(new)}$ = new speed of the queen, λ = old speed of the queen, α = a coefficient between (0,1), and γ = the value of the energy decrease. The energy and speed of the queen is updated each time that a new solution (drone) is randomly generated.

The mating flight may be considered as a set of transitions in a state space (the environment) where the queen moves between different states at variable speed and mates with the drones encountered in each state probabilistically. At the start of the flight, the queen is initialized with some energy content and returns to her nest when the energy is within some threshold from zero or when her spermatheca is full. It might be that the energy is zero but the spermatheca is not full yet. In this situation the queen uses the trial solutions to fill the spermatheca.

The queen starts breeding after completing the mating flight. A queen is selected in proportion to her fitness and inseminated with a randomly selected sperm from her spermatheca.

12.5.2 Trial Solutions

Trial solutions are used for local search with the present population. It was previously stated that mating selects drones that mate with the queen and generate new solutions. For this purpose the queen selects S_c drones for crossover. Whenever the improvement of the queen exceeds a predefined threshold, S_c trial solutions become drones and saved in the queen's spermatheca. In this case the mating flight is not carried out, and, instead, the queen's spermatheca is filled with the best solutions of the present population. However, whenever the improvement of the queen is less than a predefined threshold, then a mating flight is carried out to escape from entrapment in a local optimum. The difference between best solutions (queens) of successive iterations is evaluated as follows:

$$\varepsilon = \left| F\left(X_q^{(t)}\right) - F\left(X_q^{(t-1)}\right) \right| \tag{12.6}$$

in which $F(X_q^{(t)})$ = the queen's fitness value in iteration t and $F(X_q^{(t-1)})$ = the queen's fitness value in iteration $t-1$. If ε is less than a predefined threshold like θ, then drones are selected by mating flight in the decision space in iteration $t+1$, and randomly generated solutions fill the queen's spermatheca.

It may happen that in a mating flight, the energy of the queen is used up but the queen's spermatheca is not yet filled. In this instance, the queen's spermatheca is also filled with the best trial solutions.

12.6 Brood (New Solution) Production

New broods are produced by combining some of the queen's genes with existing genes in the sperm bag. Broods are generated by means of genetic operators including crossover and mutation operators. Crossover replaces

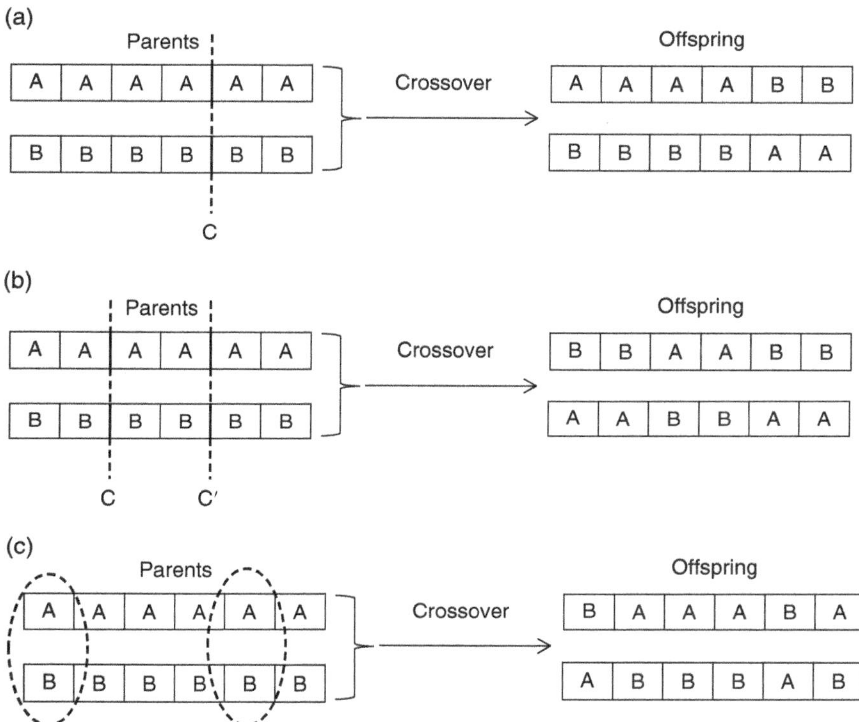

Figure 12.3 Different crossover approaches: (a) one-point crossover, (b) two-point crossover, and (c) uniform crossover.

some of the queen's genes with genes of drones memorized in the spermatheca. The crossover operator generates new offspring by exchanging some genes between the queen and the drones. Thus, crossover involves the exchange of decision variables between solutions. Goldberg (1989) and Michalewicz (1996) have described several methods of crossover including (1) one-point crossover, (2) two-point crossover, and (3) uniform crossover. Crossover occurs between two solutions. Figure 12.3 illustrates different types of crossover.

One-point crossover selects randomly a crossover point. The genes of parents placed adjacent to the crossover point generate a pair of offspring genes as shown in Figure 12.3a by interchanging the position of the parents' genes. Two-point crossover selects randomly two crossover points. The genes of the parents located between the crossover points are replicated in the offspring as shown in Figure 12.3b. The genes of the parents not placed between the crossover points are transposed in the offspring (see Figure 12.3b). Uniform crossover selects randomly parents' genes, and the offspring's genes are transposed correspondingly as shown in Figure 12.3c.

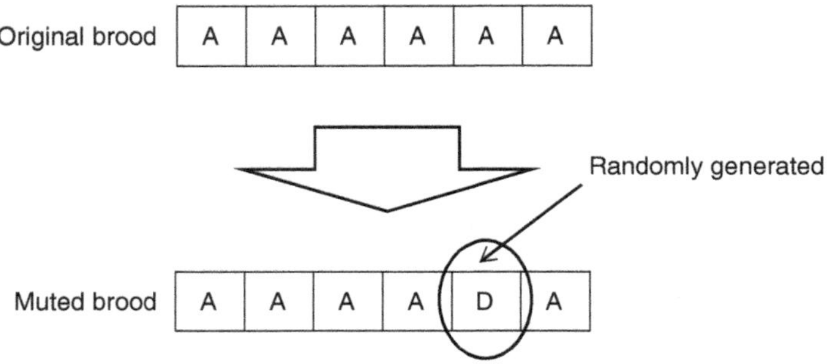

Figure 12.4 Performance of mutation operator.

A crossover point is an integer random number in the range [1,*N*]. For example, if $X = (x_1, x_2, \dots, x_N)$ denotes the queen, $X' = (x'_1, x'_2, \dots, x'_N)$ denotes a drone, and the index *C* indicates a crossover point in one-point crossover, then two children are generated as follows (see Figure 12.3a):

$$X_1^{(new)} = (x_1, x_2, \dots, x_c, x'_{c+1}, x'_{c+2}, \dots, x'_N) \tag{12.7}$$

$$X_2^{(new)} = (x'_1, x'_2, \dots, x'_c, x_{c+1}, x_{c+2}, \dots, x_N) \tag{12.8}$$

in which $X_1^{(new)}$ and $X_2^{(new)}$ = newly generated solutions by one-point crossover operator.

The mutation operator replaces one or more decision variables of the current solution with random values, and the other values of the decision variable remain unchanged. Figure 12.4 illustrates the mutation operator.

One of the methods of mutation is called uniform mutation. Uniform mutation randomly generates a value that is within the feasible range of values to replace the value of a gene. Uniform mutation prescribes that if the *i*th decision variable (x_i) of a new solution $X = (x_1, x_2, \dots, x_i, \dots, x_N)$ and is selected for mutation, then the muted brood would be $X = (x_1, x_2, \dots, x'_i, \dots, x_N)$ whereby x'_i is evaluated as follows:

$$x'_i = Rnd\left(x_i^{(L)}, x_i^{(U)}\right) \tag{12.9}$$

in which x'_i = the new value of x_i that is selected for mutation, $x_i^{(U)}$ = the upper bound of the *i*th decision variable, $x_i^{(L)}$ = the lower bound of the *i*th decision variable, and *Rnd*(*a*,*b*) = a random value from the range [*a*,*b*].

Mutation is done probabilistically. In fact, a mutation probability (P_M) is specified that permits random mutations to be made to individual genes. The mutation operator is implemented by generating a random number *Rand* in the range [0,1] for each decision variable of a new solution. If *Rand* is less than P_M, that decision variable is muted; otherwise it is remained without change.

12.7 Improving Broods (New Solutions) by Workers

The workers who feed the queen and broods are symbols of meta-heuristic functions that improve solutions. Workers that are used to improve the brood's genotype represent a set of different heuristics for local searching. In the brood caring stage of the HBMO, an attempt is made to improve the generated broods using heuristic functions. For this purpose, a heuristic function was introduced by Solgi et al. (2016a). The introduced heuristic function replaces the value of some genes of a brood with new ones that are randomly generated based on the value of the corresponding genes that belong to the brood, the queen of the previous iteration, and the queen of the present population. Let $X = (x_1, \ldots, x_n)$ denote a brood, $Y = (y_1, \ldots, y_n)$ denote the best solution in the present iteration, $Y' = (y'_1, \ldots, y'_n)$ denote the best solution in the previous iteration, and the component x_i from brood X is randomly chosen for substitution, where $X^{(new)} = (x_1, \ldots, x'_i, \ldots, x_n)$ denotes the brood after brood caring. x'_i is evaluated as follows:

$$x'_i = \phi(\delta)^2 \times G\left(y_i, \phi(\delta)\right) + \left(1 - \phi(\delta)^2\right) \times \left\{\phi(\sigma)^2 \times G\left(x_i, \phi(\sigma)\right) + \left(1 - \phi(\sigma)^2\right) \times x_i\right\} \tag{12.10}$$

where

$$G(a,b) = \frac{1+b}{2} \times Rnd\left(a, x_i^{(U)}\right) + \frac{1-b}{2} \times Rnd\left(x_i^{(L)}, a\right) \tag{12.11}$$

$$\sigma = y_i - x_i \tag{12.12}$$

$$\delta = y_i - y'_i \tag{12.13}$$

in which x_i = the value of the brood's ith component before substitution, x'_i = the value of the brood's ith component after substitution, y_i = the value of the best solution's ith component in the present iteration, y'_i = the value of the best solution's ith component in the previous iteration, and $\phi(a)$ = returns the sign of the number a (sign function) that can be equal to 1, −1, or 0. Thereafter, the functions $\phi(a)$ and $G(a,b)$ are evaluated and are substituted in Equation (12.10) as follows:

$$x'_i = \begin{cases} Rnd\left(y_i, x_i^{(U)}\right) & if\ \delta > 0 \\ Rnd\left(x_i^{(L)}, y_i\right) & if\ \delta < 0 \\ Rnd\left(x_i, x_i^{(U)}\right) & if\ \delta = 0\ and\ \sigma > 0 \\ Rnd\left(x_i^{(L)}, x_i\right) & if\ \delta = 0\ and\ \sigma < 0 \\ x_i & if\ \delta = 0\ and\ \sigma = 0 \end{cases} \tag{12.14}$$

According to Equations (12.14) if the gene of the best solution (y_i) in the present iteration is larger than that of the best solution in the previous iteration (y_i'), a random value between y_i and $x_i^{(U)}$ replaces the gene of the brood. Conversely, if y_i is less than y_i', x_i' is made equal to a random value between $x_i^{(L)}$ and y_i. When y_i and y_i' are the same ($\delta = 0$), x_i' is determined based on the result of the comparison between the present value of the brood's gene and that of the corresponding gene of the best solution in the present iteration (y_i). If y_i is larger than x_i, x_i' is made equal to a random value between x_i and $x_i^{(U)}$. If y_i is less than x_i, x_i' is made equal to a random value between $x_i^{(L)}$ and x_i. Otherwise, if y_i, y_i', and x_i are the same, the value of the brood's gene is not changed.

12.8 Termination Criteria

The termination criterion determines when to terminate the algorithm. Selecting a good termination criterion has an important role in the correct convergence of the algorithm. The number of iterations, the amount of improvement of the fitness function between consecutive iterations, and the run time are common termination criteria for the HBMO.

12.9 User-Defined Parameters of the HBMO

The size of the population (M), the size of the queen's spermatheca (S_c), θ, the value of the queen's energy decrease (γ), the reduction factor of the queen's speed (α), and termination criteria are user-defined parameters of the HBMO. A good choice of the parameters is related to the decision space of a particular problem, and usually the optimal parameter setting for one problem is of limited utility for any other problem. The difficulty associated with adjusting the values of the parameters is that the decision space is usually not well known. Consequently, determining a good parameter setting often requires the execution of trial-and-error experiments. A reasonable method for finding appropriate values for the parameters is performing sensitivity analysis. This is accomplished by choosing combinations of parameters, and the algorithm is run several times for each combination. Comparison of the results from many runs helps in determining appropriate parameter values.

12.10 Pseudocode of the HBMO

```
Begin
  Input the parameters of the algorithm and initial data
  Let θ = the threshold of applying mating flight, ψ =
    energy of the queen, and M = the population size
  Generate M initial possible solutions randomly
```

```
  While (the termination criteria are not satisfied)
     Mark the best solution out as the queen and other
       solutions as trial solutions
     Evaluate the difference between the queen of
       present and previous iteration (ε)
     If ε > θ then
         Fill the queen's spermatheca with best trial
           solutions as drones
     Otherwise
         Generate energy and speed of the queen randomly
         While (ψ > 0)
           If the queen's spermatheca is not full
                Generate a solution randomly
                If the fitness value of the generated
                 solution is better than the queen
                    Add the generated solution to the
                      queen's spermatheca
                    Update the queen's speed and energy
                Otherwise
                    Evaluate ζ(X_q, X_d) and generate Prob
                      randomly
                    If ζ(X_q, X_d) > Prob
                        Add the generated solution to
                          the queen's spermatheca
                        Update the queen's speed and
                          energy
                    End if
                End if
           End if
         End while
         If the queen's spermatheca is not full
           Fill the queen's spermatheca with trial
           solutions
         End if
     End if
     Generate broods (new solutions) by crossover and
       mutation
     Improve the broods (new solutions) by workers
     Replace all trial solutions with new generated
       solutions
     Empty the queen's spermatheca
  End while
  Report the best solution
End
```

12.11 Conclusion

This chapter described the HBMO algorithm. It contains a literature review of the HBMO and its applications. The mathematical statement of the HBMO algorithm was mapped to the basic functioning of honey-bee colonies. A pseudocode of the HBMO was presented.

References

Afshar, A., Bozorg-Haddad, O., Mariño, M. A., and Adams, B. J. (2007). "Honey-bee mating optimization (HBMO) algorithm for optimal reservoir operation." Journal of Franklin Institute, 344(5), 452–462.

Afshar, A., Shafii, M., and Bozorg-Haddad, O. (2011). "Optimizing multi-reservoir operation rules: An improved HBMO approach." Journal of Hydroinformatics, 13(1), 121–139.

Bozorg-Haddad, O. and Mariño, M. A. (2007). "Dynamic penalty function as a strategy in solving water resources combinatorial optimization problems with honey-bee mating optimization (HBMO) algorithm." Journal of Hydroinformatics, 9(3), 233–250.

Bozorg-Haddad, O. and Mariño, M. A. (2008). "Honey-bee mating optimization (HBMO) algorithm in deriving optimal operation rules for reservoirs." Journal of Hydroinformatics, 10(3), 257–264.

Bozorg-Haddad, O. and Mariño, M. A. (2011). "Optimum operation of wells in coastal aquifers." Proceedings of the Institution of Civil Engineers: Water Management, 164(3), 135–146.

Bozorg-Haddad, O., Adams, B. J., and Mariño, M. A. (2008a). "Optimum rehabilitation strategy of water distribution systems using the HBMO algorithm." Journal of Water Supply: Research and Technology, 57(5), 337–350.

Bozorg-Haddad, O., Afshar, A., and Mariño, M. A. (2006). "Honey-bees mating optimization (HBMO) algorithm: A new heuristic approach for water resources optimization." Water Resources Management, 20(5), 661–680.

Bozorg-Haddad, O., Afshar, A., and Mariño, M. A. (2008b). "Design-operation of multi-hydropower reservoirs: HBMO approach." Journal of Water Resources Management, 22 (12), 1709–1722.

Bozorg-Haddad, O., Afshar, A., and Mariño, M. A. (2009). "Optimization of non-convex water resource problems by honey-bee mating optimization (HBMO) algorithm." Engineering Computations, 26(3), 267–280.

Bozorg-Haddad, O., Afshar, A., and Mariño, M. A. (2010a). "Multireservoir optimization in discrete and continuous domains." Proceeding of the ICE: Water Management, 164(2), 57–72.

Bozorg-Haddad, O., Ghajarnia, N., Solgi, M., Loáiciga, H. A., and Mariño, M. A. (2016a). "A DSS-based honeybee mating optimization (HBMO) algorithm for single- and multi-objective design of water distribution networks." In: Yang, X.-S., Bekdas, G., and Nigdeli, S. M. (Eds.), Metaheuristics and optimization in civil engineering, Modeling and optimization in science and technologies, Vol. 7, Springer International Publishing, Cham, 199–233.

Bozorg-Haddad, O., Ghajarnia, N., Solgi, M., Loáiciga, H. A., and Mariño, M. A. (2016b). "Multi-objective design of water distribution systems based on the fuzzy reliability index." Journal of Water Supply: Research and Technology-Aqua, 66(1), 36–48.

Bozorg-Haddad, O., Hoseini-Ghafari, S., Solgi, M., and Loáiciga, H. A. (2016c). "Intermittent urban water supply with protection of consumers' welfare." Journal of Pipeline Systems Engineering and Practice, 7(3), 04016002.

Bozorg-Haddad, O., Mirmomeni, M., and Mariño, M. A. (2010b). "Optimal design of stepped spillways using the HBMO algorithm." Civil Engineering and Environmental Systems, 27(1), 81–94.

Bozorg-Haddad, O., Mirmomeni, M., Zarezadeh Mehrizi, M., and Mariño, M. A. (2010c). "Finding the shortest path with honey-bee mating optimization algorithm in project management problems with constrained/unconstrained resources." Computational Optimization and Application, 47(1), 97–128.

Bozorg-Haddad, O., Moradi-Jalal, M., and Mariño, M. A. (2011). "Design–operation optimisation of run-of-river power plants." Proceedings of the Institution of Civil Engineers: Water Management, 164(9), 463–475.

Ghajarnia, M., Bozorg-Haddad, O., and Mariño, M. A. (2011). "Performance of a novel hybrid algorithm in the design of water networks." Proceedings of the Institution of Civil Engineers: Water Management, 164(4), 173–191.

Ghajarnia, N., Bozorg-Haddad, O., and Mariño, M. A. (2009). "Reliability based design of water distribution network (WDN) considering the reliability of nodal pressure." World Environmental and Water Resources Congress 2009: Great Rivers, Kansas City, MO, May 17–21, Reston, VA: American Society of Civil Engineers (ASCE), 5717–5725.

Goldberg, D. E. (1989). "Genetic algorithms in search, optimization and machine learning." Addison-Wesley Longman, Boston.

Horng M. H., Jiang, T. W., and Chen, J. Y. (2009). "Multilevel minimum cross entropy threshold selection based on honey bee mating optimization." Proceedings of the International Multi Conference of Engineers and Computer Scientists 2009, Hong Kong, China, March 18–20, Hong Kong: International Association of Engineers.

Jahanshahi, G. and Bozorg-Haddad, O. (2008). "Honey-bee mating optimization (HBMO) algorithm for optimal design of water distribution systems." World Environmental and Water Resources Congress, Honolulu, HI, May 12–16, Reston, VA: American Society of Civil Engineers (ASCE).

Karimi, S., Mostoufi, N., and Soutodeh-Gharebagh, R. (2013). "Evaluating performance of honey bee mating optimization." Journal of Optimization Theory and Applications, 160(3), 1020–1026.

Marinakis, Y., Marinaki, M., and Matsatsinis, N. (2008a). "Honey bees mating optimization for the location routing problem." 2008 IEEE International Engineering Management Conference (IEMC-Europe 2008), Estoril, Portugal, June 28–30, Piscataway, NJ: Institute of Electrical and Electronics Engineers (IEEE).

Marinakis, Y., Marinaki, M., and Matsatsinis, N. (2008b). "A hybrid clustering algorithm based on honey bees mating optimization and greedy randomized adaptive search procedure." In: Maniezzo, V., Battiti, R., and Watson, J. P. (Eds.), Learning and intelligent optimization, Lecture notes in computer science, Vol. 5313, Springer, Berlin, Heidelberg, 138–152.

Michalewicz, Z. (1996). "Genetic algorithms + data structures = evolution programs." Springer-Verlag, Berlin, Heidelberg.

Michener, C. D. and Grimaldi, D. A. (1988). "The oldest fossil bee: Apoid history, evolutionary stasis, and antiquity of social behavior." Proceeding of the National Academy of Sciences of the United States of America, 85(17), 6424–6426.

Moritz, R. F. A. and Southwick, E. E. (1992). "Bees as superorganisms." Springer-Verlag, Berlin, Heidelberg.

Niknam, T. (2009). "An efficient hybrid evolutionary algorithm based on PSO and HBMO algorithms for multi-objective distribution feeder reconfiguration." Energy Conversion and Management, 5(8), 2074–2082.

Sabbaghpour, S., Naghashzadehgan, M., Javaherdeh, K., and Bozorg-Haddad, O. (2012). "HBMO algorithm for calibrating water distribution network of Langarud city." Water Science and Technology, 65(9), 1564–1569.

Solgi, M., Bozorg-Haddad, O., and Loáiciga, H. A. (2016a). "The enhanced honey-bee mating optimization algorithm for water resources optimization." Water Resources Management, 31(3), 885–901.

Solgi, M., Bozorg-Haddad, O., Seifollahi-Aghmiuni, S., Ghasemi-Abiazani, P., and Loáiciga, H. A. (2016b). "Optimal operation of water distribution networks under water shortage considering water quality." Journal of Pipeline Systems Engineering and Practice, 7(3), 04016005.

Solgi, M., Bozorg-Haddad, O., Seifollahi-Aghmiuni, S., and Loáiciga, H. A. (2015). "Intermittent operation of water distribution networks considering equanimity and justice principles." Journal of Pipeline Systems Engineering and Practice, 6(4), 04015004.

Soltanjalili, M. J., Bozorg-Haddad, O., and Mariño, M. A. (2011). "The effect of considering breakage level one in design of water distribution networks." Water Resources Management, 25(1), 311–337.

Soltanjalili, M. J., Bozorg-Haddad, O., and Mariño, M. A. (2013a). "Operating water distribution networks during water shortage conditions using hedging and intermittent water supply concepts." Journal of Water Resources Planning and Management, 139(6), 644–659.

Soltanjalili, M. J., Bozorg-Haddad, O., Seifollahi-Aghmiuni, S., and Mariño, M. A. (2013b). "Water distribution network simulation by optimization approaches." Water Science and Technology: Water Supply, 13(4), 1063–1079.

13

Invasive Weed Optimization

Summary

This chapter describes the invasive weed optimization (IWO) algorithm, which mimics weed's adaptive patterns. This chapter contains a literature review of the IWO, an overview of weeds' biology, a description of the mapping of the IWO algorithm to weeds' biology, a thorough explanation of the steps of the IWO algorithm, and a pseudocode of the IWO algorithm.

13.1 Introduction

Invasive weed optimization (IWO) was developed by Mehrabian and Lucas (2006). They solved two engineering problems and compared the results with other algorithms including the genetic algorithm (GA), particle swarm optimization (PSO) algorithm, the shuffled frog leading algorithm (SFLA), and the simulated annealing (SA) algorithm. The results showed a relatively superior performance by the IWO. The IWO has been implemented in a variety of engineering optimization problems. Mehrabian and Yousefi-Koma (2007) applied the IWO to optimize the location of piezoelectric actuators on a smart fin. Mallahzadeh et al. (2008) tested the flexibility, effectiveness, and efficiency of the IWO in optimizing a linear array of antenna and compared the computed results with those of the PSO algorithm. Sahraei-Ardakani et al. (2008) implemented IWO to optimize the generation of electricity. Roshanaei et al. (2009) applied the IWO to optimize uniform linear array (ULA) used in wireless networks, such as commercial cellular systems, and compared their results with those from the GA and least mean squares (LMS). Mallahzadeh et al. (2009) applied the IWO to design vertical antenna elements with maximal efficiency. Krishnanand et al. (2009) compared the effectiveness of the IWO, GA, PSO algorithm, artificial bee colony (ABC), and artificial immune (AI) by solving five basic standard mathematical problems with multivariate functions. Zhang et al. (2010) used

Meta-Heuristic and Evolutionary Algorithms for Engineering Optimization,
First Edition. Omid Bozorg-Haddad, Mohammad Solgi, and Hugo A. Loáiciga.

heuristic algorithm concepts for developing the IWO. They introduced the IWO with crossover function and tested the new algorithm on standard mathematical problems and compared the results of the developed IWO with those of the standard IWO and PSO. Sharma et al. (2011) applied the IWO to schedule dynamic economic dispatching (DED). Their results showed that the IWO algorithms reduced production costs relative to those obtained with the PSO and AI algorithms and the differential evolution (DE). Jayabarathi et al. (2012) employed the IWO for solving economic dispatch problems. Kostrzewa and Josiński (2012) developed a new version of the IWO and tested their algorithm on several standard mathematical problems. Abu-Al-Nadi et al. (2013) implemented the IWO for model order reduction (MOR) in linear multiple-input–multiple-output (MIMO) systems. Sang and Pan (2013) introduced the effective discrete IWO (DIWO) to solve the problem of flow shop scheduling with average stored buffers and compared their results with the hybrid GA (HGA), hybrid PSO algorithm (HPSO), and the hybrid discrete differential evolution algorithm (HDDE). Saravanan et al. (2014) applied the IWO to solve the unit commitment (UC) problem for minimizing the total costs of generating electricity. They compared their results with those calculated with the GA, SFLA, PSO, Lagrangian relaxation (LR), and bacterial foraging (BF) algorithms. Barisal and Prusty (2015) applied the IWO to solve economic problems on a large scale with the aim of minimizing the costs of production and transfer of goods subject to restrictions on production, market demand, and the damage caused to goods during transportation and to alleviate other calamities. Asgari et al. (2015) presented a modified IWO as weed optimization algorithm (WOA) to optimal reservoir operation. Hamedi et al. (2016) applied the WOA for parameter estimation of hydrologic flood-routing models.

13.2 Mapping Invasive Weed Optimization (IWO) to Weeds' Biology

Weeds grow spontaneously and compete with other vegetation. A plant is called weed if in any specified geographical area, its population grows entirely or predominantly in conditions markedly disturbed by man. Weeds are agricultural pests. They can easily adapt to almost any environment and new conditions. It is a common belief in agronomy that "The Weeds Always Win." Weeds may reproduce with or without sexual cells depending on the type of the plant. In sexual reproduction eggs are fertilized by pollen and form seeds in a parent plant. Several factors such as wind, water, and animals distribute seeds. When seeds find suitable place to thrive, they grow to adult plants while in interaction with other neighboring plants. They turn to flowering plants and produce seeds at the final stage of their life. The weed biomass produced becomes limited by the availability of resources so that the yield per unit area becomes independent of density. The stress of density increases the risk of mortality to whole plants and their parts, and the rate of death becomes a function of the growth rate of the survivors. Thus, birth, growth, and reproduction of plants are influenced by population density. There are three main components of compatibility in the community. They are (1) reproduction, (2) struggle for survival with competitors, and (3) avoidance of predators. Any weed colony tries to improve its compatibility to achieve a longer life. The study of population biology seeks to unravel the factors that are important for weed survival and reproduction. One of the factors is called r-selection, which implies "live fast, reproduce quick, die young." r-selection enhances the chances to succeed in unstable and unpredictable environments, where ability to reproduce rapidly and opportunistically is at a premium and where there is little value in adaptations to succeed in competition. A variety of qualities are thought to be favored by r-selection, including high fecundity, small size, and adaptations for long-distance dispersal. On the other hand, K-selection is tantamount to "live slow, reproduce slow, die old." Selection for the qualities is needed to succeed in stable, predictable environments where there is likely to be heavy competition for limited resources between individuals well equipped to compete when the population size is close to the maximum that the habitat can bear. A variety of qualities are thought to be favored by K-selection, including large size, long life, and small numbers of intensively cared-for offspring, in contrast with r-selection (Mehrabian and Lucas, 2006). It is customary to emphasize that r-selection and K-selection are the extremes of a continuum, most real cases lying somewhere in between.

IWO represents a solution with a plant whose location in an N-dimensional space is a decision variable. A bunch of plants constitutes a colony. In nature each weed, based on its quality in the colony, produces seeds that spread randomly in the environment, grows, and eventually generates new seeds. Therefore, each plant generates a specified number of new seeds according to its fitness value. Each seed is known as a new solution. If the maximum number of plants in a colony is reached, competition for survival starts between weeds so that in each stage weeds with lower quality (less fitness value) are removed. The remaining weeds as mother plants spread new seeds. This process continues to produce weeds of the highest quality (the best fitness value). Table 13.1 shows the characteristics of the IWO.

The IWO starts the optimization process by randomly generating a set of weeds, each of which is a solution of the optimization problem (see Section 2.6). After evaluating the fitness function for all solutions, the number of seeds for each weed (solution) is estimated based on its fitness value. All weeds (solutions) generate seeds (new solutions). Solutions with low fitness are eliminated until the number of solutions equals the capacity of colony whenever the number of solutions exceeds a threshold. Improved, new solutions are generated by remaining solutions for as long as the termination criteria are not satisfied. Figure 13.1 depicts the flowchart of the IWO.

Table 13.1 The characteristics of the IWO.

General algorithm (see Section 2.13)	Invasive weed optimization
Decision variable	Weed's location in each dimension
Solution	Weed (position)
Old solution	Mother plant
New solution	Seed
Best solution	–
Fitness function	Quality of the plant
Initial solution	Random weed
Selection	Competition for survival
Process of generating new solution	Spreading seeds

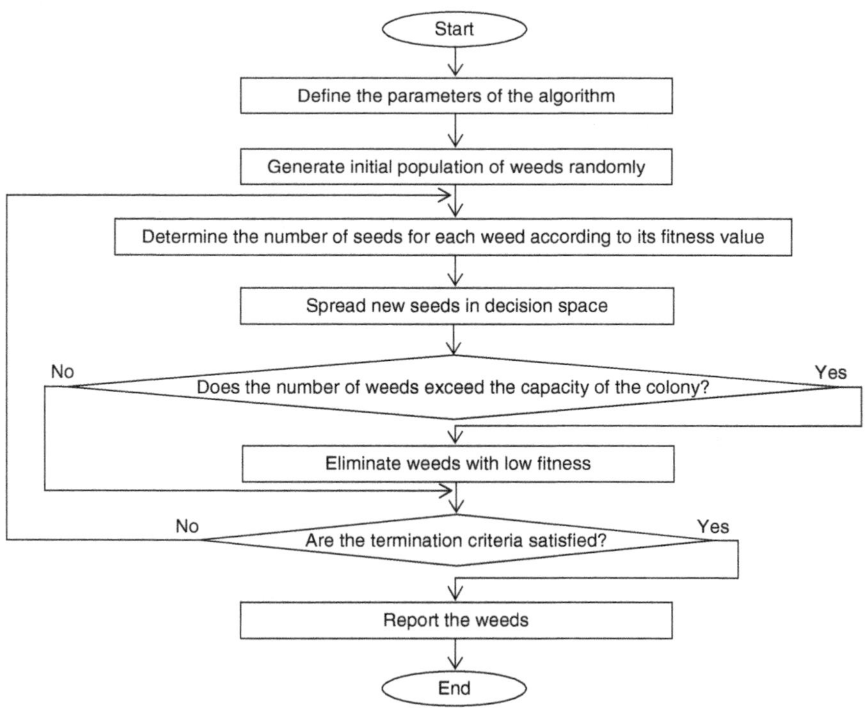

Figure 13.1 The flowchart of the IWO.

13.3 Creating an Initial Population

IWO calls each possible solution of the optimization problem a weed. A weed's location in any dimension of an N-dimensional space is a decision variable of the optimization problem, and a weed is represented as an array of size $1 \times N$ that describes a weed's location. This array is defined as follows:

$$Weed = X = \left(x_1, x_2, \ldots, x_i, \ldots, x_N\right) \tag{13.1}$$

where X = a solution of the optimization problem, x_i = ith decision variable of the solution X, and N = number of decision variables. The decision variable values $(x_1, x_2, x_3, \ldots, x_N)$ are represented as floating point number (real values) or as a predefined set for continuous and discrete problems, respectively.

The IWO starts by randomly generating a matrix of size $M \times N$ (see Section 2.6), where M and N are the size of population and the number of decision variables, respectively. Hence, the matrix of solutions that is generated randomly is as follows (rows and columns are the number of weeds and the number of decision variables, respectively):

$$Population = \begin{bmatrix} X_1 \\ X_2 \\ \vdots \\ X_j \\ \vdots \\ X_M \end{bmatrix} = \begin{bmatrix} x_{1,1} & x_{1,2} & \cdots & x_{1,i} & \cdots & x_{1,N} \\ x_{2,1} & x_{2,2} & \cdots & x_{2,i} & \cdots & x_{2,N} \\ & & & \vdots & & \\ x_{j,1} & x_{j,2} & \cdots & x_{j,i} & \cdots & x_{j,N} \\ & & & \vdots & & \\ x_{M,1} & x_{M,2} & \cdots & x_{M,i} & \cdots & x_{M,N} \end{bmatrix} \tag{13.2}$$

in which X_j = jth solution, $x_{j,i}$ = the ith decision variable of the jth solution, and M = size of the initial population of weeds.

13.4 Reproduction

During the reproductions stage, weeds are allowed to produce seeds according to their fitness values and the maximum and minimum allowed numbers of produced seeds (λ_{max}), (λ_{min}), respectively. The solution with the worst fitness value generates λ_{min} new solutions, while the best solution generates λ_{max} new solutions. Other solutions generate new solutions according to their fitness function between these two limiting values. The number of seeds for each solution is evaluated as follows:

$$\mu_j = \frac{\left(\lambda_{max} - \lambda_{min}\right)}{|Best - Worst|} \times F(X_j), \quad j = 1, 2, \ldots, M \tag{13.3}$$

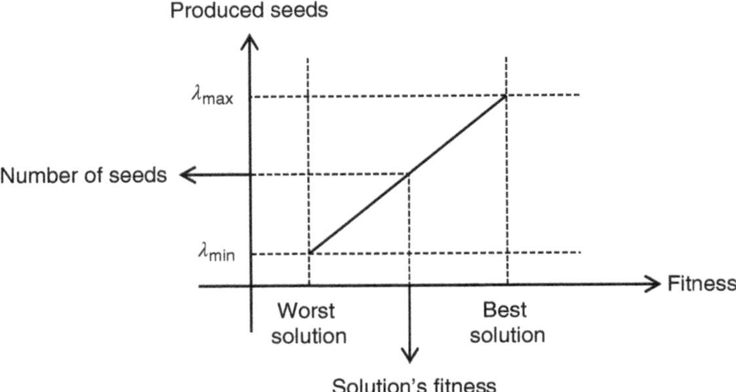

Figure 13.2 Number of seeds for each weed with respect to the fitness value.

in which μ_j= the number of new solutions generated by the *j*th solution, $F(X_j)$ = fitness value of the *j*th solution, *Best* = the fitness value of the best solution, *Worst* = the fitness value of the worst solution. and λ_{max} and λ_{min} = the maximum and minimum number of produced seeds, respectively, which are user-defined parameters. Seed production is illustrated in Figure 13.2 approximately as a linear function.

The reproduction stage adds an important advantage to the IWO algorithm. Evolutionary algorithms have population members that range from appropriate solutions to inappropriate ones. Appropriate solutions have a higher probability of reproduction than inappropriate ones, but there is always the possibility that population elements that seem inappropriate at each stage contain important information that even suitable plants lack. It is possible that some individuals with lower fitness value carry useful information during the evolution process. Therefore, the system can reach the optimal point more easily if it is possible to "cross" an infeasible region. It is therefore probable that with a suitable reproduction, inappropriate plants survive an unsuitable environment and find a hidden suitable environment. This phenomenon is observed in nature.

13.5 The Spread of Seeds

Adoption and randomness are introduced in the IWO algorithm by the spread of seeds. The produced seeds spread randomly with a normal distribution that has zero mean in an *N*-dimensional space. Therefore, new plants spread randomly around the mother plants, but their standard deviation is variable.

The standard deviation is reduced from the initial predetermined value (maximum) to a final predetermined value (minimum) as follows:

$$\sigma_t = \frac{(T-t)^\beta}{(T)^\beta}\left(\sigma_{initial} - \sigma_{final}\right) + \sigma_{final} \tag{13.4}$$

where σ_t = the standard deviation of the current iteration t, T = the maximum number of iterations (reproduction stages), t = the current iteration, $\sigma_{initial}$ = the initial standard deviation, σ_{final} = the final standard deviation, and β = a nonlinear modulus (called nonlinear modulation index) selected by the user as parameter of the algorithm.

New solutions are generated after evaluating the standard division as follows:

$$X_r^{(new)} = Mrand(0,\sigma_t) + X_j, \quad r = 1,2,\ldots\mu_j, \quad j = 1,2,\ldots,M \tag{13.5}$$

in which $X_r^{(new)}$ = the new solution rth on the basis of solution jth of the population, $Mrand(0, \sigma_t)$ = a matrix of random values by mean 0 and standard division σ_t with size $1 \times N$, and X_j = jth solution of the population.

The algorithm localizes search during its progress. At the beginning of the algorithm, the standard deviation is high; while the algorithm proceeds, the standard deviation is reduced by Equation (13.4). In Equation (13.5) a low standard deviation makes new solutions to be close to the mother plant, while a high standard deviation makes new solutions spread out over a wider range. The probability of placing a seed far from its mother plant in the beginning of the algorithm is high, and it decreases during later stages of the algorithm when the number of appropriate plants increases. In this manner the probability of dropping a seed in a distant area decreases nonlinearly at each time step, which results in grouping fitter plants and elimination of inappropriate plants. This represents transformation from r-selection to K-selection.

13.6 Eliminating Weeds with Low Fitness

A plant that does not produce seeds becomes extinct. The number of plants increases exponentially if all the plants produce seeds and the seeds grow. Therefore, a competitive process is necessary to limit and remove some of the existing plants. After several reproductions the number of plants in the colony reaches its maximum (M_{max}). At this juncture the process of omitting unsuitable plants starts and is repeated until the end of the algorithm.

The elimination scheme allows each weed to produce seeds according to reproduction rules whenever the current population size (M) reaches M_{max}. Offspring (new solutions) are ranked together with their parents (as a colony of weeds). Next, weeds with low fitness are eliminated until the number of weeds is equal to the maximum allowable population in the colony. In this manner plants with lower fitness have a chance to reproduce, and if their offspring have a good fitness in the colony, then they can survive.

13.7 Termination Criteria

The termination criterion ends the execution of the IWO algorithm. Selecting a good termination criterion has an important role in the correct convergence of the algorithm. The number of iterations, the amount of improvement of the solution between consecutive iterations, and the run time are common termination criteria for the IWO.

13.8 User-Defined Parameters of the IWO

The initial size of population (M), the maximum size of the population (M_{max}), the maximum number of produced seeds (λ_{max}), the minimum number of produced seeds (λ_{min}), the initial standard deviation ($\sigma_{initial}$), the final standard deviation (σ_{final}), the nonlinear modulation index (β), and the criterion used to decide when to terminate the algorithm are user-defined parameters of the IWO algorithm. A good choice of the parameters is related to the decision space of a particular problem, and usually the optimal parameter setting for one problem is of limited utility for any other problem. The issue of how to determine the appropriate values of these parameters is pertinent. Practice and experience with specific types of problems is valuable in this respect. A reasonable method for finding appropriate values for the parameters is performing sensitivity analysis. This is accomplished by choosing combinations of parameters and running the algorithm several times for each combination. A comparison of the results from different runs helps in determining appropriate parameter values.

13.9 Pseudocode of the IWO

```
Begin
  Input the parameters of the algorithm and initial data
  Let Mmax = the maximum population size and M = the
    current population size
  Generate M initial possible solutions randomly and
    evaluate their fitness values
  While (the termination criteria are not satisfied)
```

```
      Evaluate the standard division (σ)
      For j = 1 to M
          Determine the number of seeds (μj) for solution j
            according to its fitness value
          For r = 1 to μj
              Generate a new solution around the solution
                j using normal distribution and add it to
                the offspring population
              Evaluate fitness value of the newly generated
                solution
          Next r
      Next j
      Add offspring population to the current
        population and update M
      If M > Mmax
          Eliminate solutions with low fitness until M = Mmax
      End if
  End while
  Report the population
End
```

13.10 Conclusion

This chapter described the IWO algorithm. The IWO is a meta-heuristic optimization method inspired by weeds' ecological characteristics. This chapter presented literature review of the IWO and its applications, the weeds' biology was mapped into a mathematical statement of the IWO algorithm, each part of the IWO was explained in detail, and a pseudocode of the algorithm was presented.

References

Abu-Al-Nadi, D. I., Alsmadi, O. M., Abo-Hammour, Z. S., Hawa, M. F., and Rahhal, J. S. (2013). "Invasive weed optimization for model order reduction of linear MIMO systems." Applied Mathematical Modeling, 37(13), 4570–4577.

Asgari, H. R., Bozorg-Haddad, O., Pazoki, M., and Loáiciga, H. A. (2015). "Weed optimization algorithm for optimal reservoir operation." Journal of Irrigation and Drainage Engineering, 142(2), 04015055.

Barisal, A. K. and Prusty, R. C. (2015). "Large scale economic dispatch of power systems using oppositional invasive weed optimization." Applied Soft Computing, 29, 122–137.

Hamedi, F., Bozorg-Haddad, O., Pazoki, M., and Asgari, H. R. (2016). "Parameter estimation of extended nonlinear Muskingum models with the weed optimization algorithm." Journal of Irrigation and Drainage Engineering, 142(12), 04016059.

Jayabarathi, T., Yazdani, A., and Ramesh, V. (2012). "Application of the invasive weed optimization algorithm to economic dispatch problems." Frontiers in Energy, 6(3), 255–259.

Kostrzewa, D. and Josiński, H. (2012). "The modified IWO algorithm for optimization of numerical functions." In: Rutkowski, L., Korytkowski, M., Scherer, R., Tadeusiewicz, R., Zadeh, L. A., and Zurada, J. M. (Eds.), Swarm and evolutionary computation, Lecture notes in computer science, Vol. 7269, Springer, Berlin, Heidelberg, 267–274.

Krishnanand, K. R., Nayak, S. K., Panigrahi, B. K., and Rout, P. K. (2009). "Comparative study of five bio-inspired evolutionary optimization techniques." 2009 World Congress on Nature and Biologically Inspired Computing (NaBIC), Coimbatore, India, December 11–19, Piscataway, NJ: Institute of Electrical and Electronics Engineers (IEEE), 1231–1236.

Mallahzadeh, A. R., Es'haghi, S., and Alipour, A. (2009). "Design of an E-shaped MIMO antenna using IWO algorithm for wireless application at 5.8 GHz." Progress in Electromagnetics Research, 90(8), 187–203.

Mallahzadeh, A. R., Oraizi, H., and Davoodi-Rad, Z. (2008). "Application of the invasive weed optimization technique for antenna configurations." Progress in Electromagnetics Research, 79(8), 137–150.

Mehrabian, A. R. and Lucas, C. (2006). "A novel numerical optimization algorithm inspired from weed colonization." Ecological Informatics, 1(4), 355–366.

Mehrabian, A. R. and Yousefi-Koma, A. (2007). "Optimal positioning of piezoelectric actuators on a smart fin using bio-inspired algorithms." Aerospace Science and Technology, 11(2), 174–182.

Roshanaei, M., Lucas, C., and Mehrabian, A. R. (2009). "Adaptive beam forming using a novel numerical optimization algorithm." IET Microwaves, Antennas and Propagation, 3(5), 765–773.

Sahraei-Ardakani, M., Roshanaei, M., Rahimi-Kian, A., and Lucas, C. (2008). "A study of electricity market dynamics using invasive weed colonization optimization." 2008 IEEE Symposium on Computational Intelligence and Games, Perth, Australia, December 15–18, Piscataway, NJ: Institute of Electrical and Electronics Engineers (IEEE), 276–282.

Sang, H. Y. and Pan, Q. K. (2013). "An effective invasive weed optimization algorithm for the flow shop scheduling with intermediate buffers." 25th Chinese Control and Decision Conference (CCDC), Guiyang, China, May 25–27, Piscataway, NJ: Institute of Electrical and Electronics Engineers (IEEE), 4570–4577.

Saravanan, B., Vasudevan, E. R., and Kothari, D. P. (2014). "Unit commitment problem solution using invasive weed optimization algorithm." International Journal of Electrical Power and Energy Systems, 55, 21–28.

Sharma, R., Nayak, N., Krishnanand, K. R., and Rout, P. K. (2011). "Modified invasive weed optimization with dual mutation technique for dynamic economic dispatch." International Conference on Energy, Automation, and Signal (ICEAS), Bhubaneswar, Orissa, India, December 28–30, Piscataway, NJ: Institute of Electrical and Electronics Engineers (IEEE), 1–6.

Zhang, X., Niu, Y., Cui, G., and Wang, Y. (2010). "A modified invasive weed optimization with crossover operation." 8th World Congress on Intelligent Control and Automation (WCICA), IEEE, Jinan, China, July 7–9, Piscataway, NJ: Institute of Electrical and Electronics Engineers (IEEE), 1–11.

14

Central Force Optimization

Summary

This chapter describes the central force optimization (CFO) algorithm. The basic concepts of the CFO are issued from kinesiology in physics. The CFO resembles the motion of masses under the influence of the gravity field. One of the most important features of the CFO is that it is a deterministic method, which means that each position of a particle (called probe in this method) follows a certain path toward a solution. The following sections relate Newton's gravitational low and the CFO. The CFO algorithm is explained, and a pseudocode of the algorithm is presented.

14.1 Introduction

The central force optimization (CFO) is a search meta-heuristic method developed by Formato (2007) based on gravitational kinematics. This algorithm models the motion of airborne probes under effect of gravity and maps the equations' motion to an optimization scheme. The CFO algorithmic equations are developed for the probes' positions and the accelerations using the analogy of particle motion in a gravitational field. The CFO is deterministic, which is a variance from most other meta-heuristic algorithms. Formato (2007) assessed the performance of the CFO algorithm with recognized complex mathematical functions and electronic problems and compared the results with that of other algorithms. Formato (2010) demonstrated the good performance of the CFO algorithm in solving several different functions. Mahmoud (2011) applied the CFO to a microstrip antenna design problem. Formato (2012) employed the CFO in electronics for improving bandwidth and achieved very precise results. Also, Haghighi and Ramos (2012) applied the CFO algorithm for drinking-water networks and compared the results with previous works.

Meta-Heuristic and Evolutionary Algorithms for Engineering Optimization,
First Edition. Omid Bozorg-Haddad, Mohammad Solgi, and Hugo A. Loáiciga.

The results demonstrated that the CFO algorithm achieved solutions more rapidly. Liu and Tian (2015) developed a multi-start CFO (MCFO) and compared the performance of the MCFO with those of other algorithms.

14.2 Mapping Central Force Optimization (CFO) to Newton's Gravitational Law

The CEO is inspired by the motion of masses in a gravitational field. The gravitational force between two masses $Mass_1$ and $Mass_2$ is described by Newton's universal law of gravitation as follows:

$$Force = \gamma \frac{Mass_1 \times Mass_2}{r^2} \tag{14.1}$$

in which $Force$ = the magnitude of the force of attraction, γ = gravitational constant, $Mass_1$ and $Mass_2$ = masses that attract each other, and r = the distance between the center of masses of $Mass_1$ and $Mass_2$. According to the gravitational force, each mass such as $Mass_1$ is accelerated toward another mass $Mass_2$ with vector acceleration calculated as follows:

$$\vec{A}_1 = \gamma \frac{Mass_2 \times \hat{r}}{r^2} \tag{14.2}$$

in which $\vec{A}_1$ = acceleration vector of $Mass_1$ toward $Mass_2$ and $\hat{r}$ = a unit vector.

The gravitational force causes particles to move toward each other. The new particle's position is calculated as follows:

$$\vec{R}(t+\Delta t) = \vec{R}(t) + \vec{V}(t)\Delta t + \frac{\vec{A} \cdot \Delta t^2}{2} \tag{14.3}$$

in which $\vec{R}(t)$ = the position of particle at time t, $\vec{V}(t)$ = the velocity of particle at time t, and Δt = time interval.

The CFO has a physical base. Suppose that we want to find the largest planet in a hypothetical star system whose position is unknown. From the gravitational law described previously, it can be inferred that the largest planet has the strongest gravitational field. Therefore, if several probe satellites are spread through the star system, they gradually move along gravitational fields. After a long enough time, most of the probes probably will cluster in orbits surrounding the planet with the largest gravitational field. The CFO generalizes the equations of motion in three-dimensional physical space to seek optima in a multidimensional decision space.

CFO designates the location of each probe as a solution of the optimization problem. All particles have masses proportional to their fitness values so that the heavier the masses, the better the fitness values. According to the

Table 14.1 The characteristics of the CFO.

General algorithm (see Section 2.13)	Central force optimization
Decision variable	Position of probes in each dimension
Solution	Position of probe
Old solution	The old position of probe
New solution	The new position of probe
Best solution	–
Fitness function	Mass of probe
Initial solution	Deterministic position
Selection	–
Process of generating new solutions	Movement of probe

gravitational law, probes move toward each other with a specific velocity and acceleration. Movements of probes through the decision space produce new solutions. Table 14.1 lists the characteristics of the CFO.

The CFO starts by specifying initial probe positions deterministically as explained in the next section. Then fitness values are evaluated and the initial acceleration is assigned to each probe. The new positions of probes are computed based on the previously evaluated accelerations. Each probe must be located inside the decision space. If a probe strays outside the decision space, it is called a deviated probe, and its location is modified. The fitness values of the new locations are evaluated and new accelerations are estimated. This process is repeated until the termination criteria are satisfied. Figure 14.1 illustrates the flowchart of the CFO.

14.3 Initializing the Position of Probes

The CFO calls each possible solution of the optimization problem a probe. A probe's position in an N-dimensional optimization problem is a decision variable of the optimization problem. A probe is represented by an array of size $1 \times N$ that expresses the probe's position. This array is defined as follows:

$$Probe = X = \left(x_1, x_2, \ldots, x_i, \ldots, x_N\right) \tag{14.4}$$

where X = a solution of optimization problem, x_i = ith decision variable of solution X, and N = number of decision variables. The decision variable values $(x_1, x_2, x_3, \ldots, x_N)$ are represented as floating-point numbers (real values) or as a predefined set of values for continuous and discrete problems, respectively.

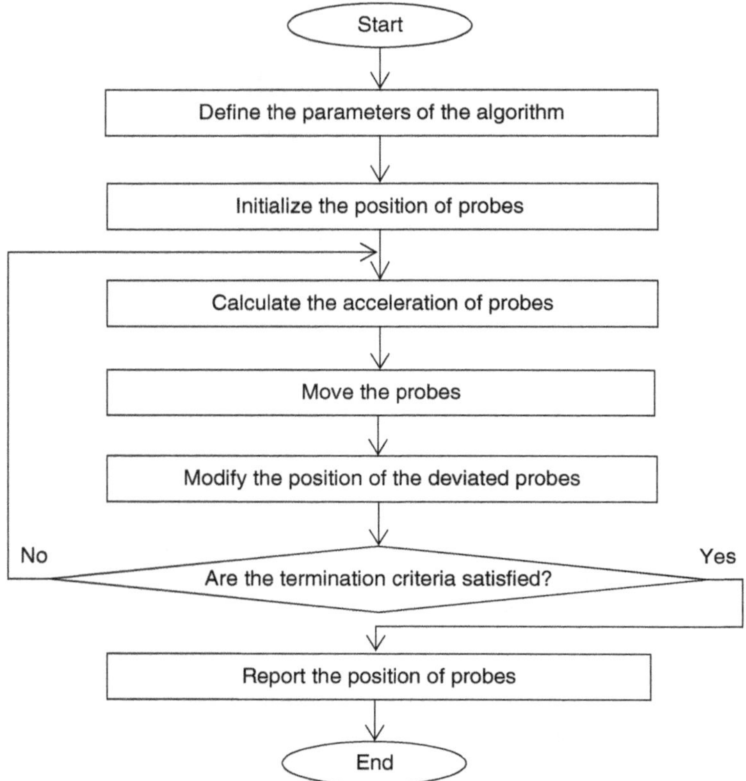

Figure 14.1 The flowchart of the CFO.

A matrix of probes of size $M \times N$ is generated (where M is the number of probes) to start the CFO algorithm. The matrix of probes is represented as follows (rows and columns denote the number of probes and the number of decision variables, respectively):

$$Population = \begin{bmatrix} X_1 \\ X_2 \\ \vdots \\ X_j \\ \vdots \\ X_M \end{bmatrix} = \begin{bmatrix} x_{1,1} & x_{1,2} & \cdots & x_{1,i} & \cdots & x_{1,N} \\ x_{2,1} & x_{2,2} & \cdots & x_{2,i} & \cdots & x_{2,N} \\ & & & \vdots & & \\ x_{j,1} & x_{j,2} & \cdots & x_{j,i} & \cdots & x_{j,N} \\ & & & \vdots & & \\ x_{M,1} & x_{M,2} & \cdots & x_{M,i} & \cdots & x_{M,N} \end{bmatrix} \tag{14.5}$$

in which X_j = jth solution, $x_{j,i}$ = ith decision variable of the jth solution, and M = population size.

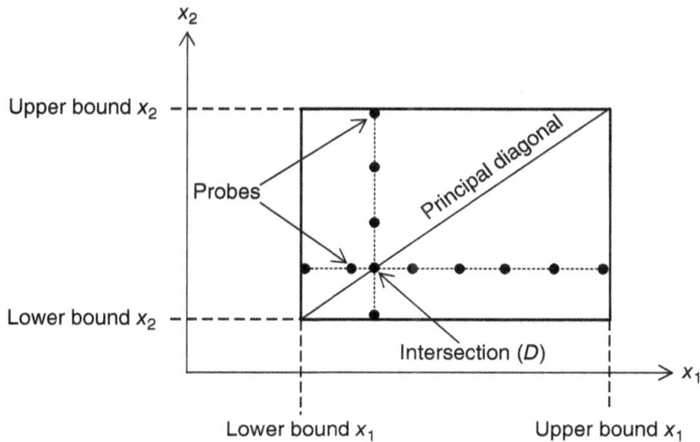

Figure 14.2 Distribution of initial probes in the CFO algorithm.

Notice that unlike most of other meta-heuristic algorithms, the initial solutions in the CFO are not generated randomly. Formato (2007) did not present a general scheme for generating initial solutions. According to Formato (2011) the initial probes are distributed on lines parallel to the coordinate axes that contain the decision space of the optimization problem. Figure 14.2 depicts distribution of the initial probes in a two-dimensional problem.

It is seen in Figure 14.2 that lines made of initial probes intersect at a point along the principal diagonal. The intersection is calculated as follows:

$$D = X_{min} + \delta\left(X_{max} - X_{min}\right) \tag{14.6}$$

$$X_{min} = \sum_{i=1}^{N} x_i^{(L)} \hat{e}_i \tag{14.7}$$

$$X_{max} = \sum_{i=1}^{N} x_i^{(U)} \hat{e}_i \tag{14.8}$$

in which D = the intersection position, X_{min} and X_{max} = diagonal's end points, δ = a parameter that determines where along the diagonal the orthogonal probe array is placed ($0 \le \delta \le 1$), $x_i^{(L)}$ = the lower bound of the ith decision variable, $x_i^{(U)}$ = the upper bound of the ith decision variable, and $\hat{e}_i$ = a matrix of size $1 \times N$ whose elements are zero except for its ith element that equals 1. Different numbers of initial probes per axis can be implemented if, for instance, equal probe spacings were desired in a space with unequal boundaries or if excluding overlapping probes is intended (Formato, 2011).

14.4 Calculation of Accelerations

According to Newton's universal law of attraction, each probe experiences an acceleration vector under the influence of the gravitational central forces generated by other probes. In a maximizing optimization problem, the "acceleration" experienced by the jth probe due to the kth probe is calculated as follows:

$$a_{j,i}^{(k)} = G \times U\left(F(X_k) - F(X_j)\right) \times \left(F(X_k) - F(X_j)\right)^{\alpha} \times \frac{x_{k,i} - x_{j,i}}{\left(x_{k,i} - x_{j,i}\right)^{\beta}} \tag{14.9}$$

$$i = 1,2,\ldots,N, \quad j = 1,2,\ldots,M$$

$$U(z) = \begin{cases} 1 & if\ z \geq 0 \\ 0 & if\ z < 0 \end{cases} \tag{14.10}$$

$$A_j^{(k)} = \left(a_{j,1}^{(k)}, a_{j,2}^{(k)}, \ldots, a_{j,i}^{(k)}, \ldots, a_{j,N}^{(k)}\right), \quad j = 1,2,\ldots,M \tag{14.11}$$

in which $a_{j,i}^{(k)}$ = the acceleration of the jth probe due to the kth probe in the ith dimension; α, β, and G = gravitational constant ($G > 0$; $\alpha > 0$ and $\beta > 0$); $F(X)$ = the fitness value of solution X; $U(z)$ = the unit step function; and $A_j^{(k)}$ = acceleration vector of the jth probe due to the kth probe.

The CFO algorithm permits assigning a gravitational acceleration different from the actual one in the physical universe. For example, the term "$(F(X_k) - F(X_j))$" in Equation (14.9) resembles mass in Equation (14.2). Probes that are located near each other in the decision space may have similar fitness values. This may lead to an excessive gravitational force on the subject probe. Therefore, it is the difference between fitness values, instead of the fitness values themselves, that is used to avoid excessive gravitational attraction by other probes that are very close to the subject probe. The fitness difference intuitively seems to be a better measure of how much gravitational influence there should be between the probe with a greater fitness and the probe with a smaller one (Formato, 2007).

Although real masses are positive, the term "$(F(X_k) - F(X_j))$" can be positive or negative depending on which objective function is greater. The unit step function is introduced to avoid the possibility of "negative" mass. In other words, using the unit step function forces the CFO to create only positive masses as observed in nature. In case negative masses are allowed, the corresponding accelerations are repulsive instead of attractive. The effect of a repulsive gravitational force is that probes move far away from large fitness values instead of being attracted toward them.

Equation (14.9) evaluates the acceleration of the jth probe toward the kth probe. Notice that other probes may attract the jth probe and also affect its acceleration. The total acceleration of the jth probe is equal to the summation of all the accelerations exerted by all other probes, and it is calculated as follows:

$$a_{j,i} = G \times \sum_{\substack{k=1 \\ k \neq j}}^{M} \left[U\left(F(X_k) - F(X_j)\right) \times \left(F(X_k) - F(X_j)\right)^{\alpha} \times \frac{x_{k,i} - x_{j,i}}{\left(x_{k,i} - x_{j,i}\right)^{\beta}} \right]$$

$$j = 1,2,\ldots,M, \quad i = 1,2,\ldots,N \tag{14.12}$$

$$A_j = \left(a_{j,1}, a_{j,2}, \ldots, a_{j,i}, \ldots, a_{j,N}\right) \tag{14.13}$$

in which $a_{j,i}$ = overall acceleration of the jth probe in the ith dimension and A_j = overall acceleration of the jth probe due to all other probes.

14.5 Movement of Probes

Probes move through the space and reach new positions in each iteration as the decision space is searched and new solutions are generated. Moving to a new position is done based on the current position of the probe, the previous velocity of the probe, and its acceleration. The new position is evaluated as follows:

$$x'_{j,i} = x_{j,i} + \left(v_{j,i} \times \psi\right) + \frac{1}{2}\left(a_{j,i} \times \psi^2\right), \quad i = 1,2,\ldots,N, \quad j = 1,2,\ldots,M \tag{14.14}$$

$$v_{j,i} = \frac{x_{j,i} - x_{j,i}^{(old)}}{\psi}, \quad i = 1,2,\ldots,N, \quad i = 1,2,\ldots,M \tag{14.15}$$

$$X_j^{(new)} = \left(x'_{j,1}, x'_{j,2}, \ldots, x'_{j,i}, \ldots, x'_{j,N}\right), \quad j = 1,2,\ldots,M \tag{14.16}$$

in which $x'_{j,i}$ = the new value of the ith decision variable of the jth solution, $X_j^{(new)}$ = the new position of the jth probe (new solution), $x_{j,i}$ = the current value of the ith decision variable of the jth solution at the present iteration, $x_{j,i}^{(old)}$ = the value of the ith decision variable of the jth solution in the previous iteration, ψ = parameter of the algorithm that resembles time interval in physics, and $v_{j,i}$ = velocity of the jth solution in the ith dimension. Formato (2007) suggested that the initial value of v and ψ be considered to be zero and one, respectively. In other words the value of v in the first iteration is equal to zero for all probes in all dimensions.

14.6 Modification of Deviated Probes

While the algorithm progresses, some probes may move to a position outside the decision space. A probe that strays outside the decision space is called a deviated probe and its location has to be modified. The method for such modification is central to the proper convergence of the CFO.

There are many possible approaches to returning deviated probes to the feasible space. One method is returning the probe to a specific point such as its starting point or its last position. However, Formato (2007) stated that this method does not work well. Another method is that any probe outside the decision space is returned to the midpoint between its starting position and the minimum or maximum value of the coordinate lying outside the allowable range. Another possibility is to randomly reposition deviated probes. This method is a simple approach because it can utilize the compiler's built-in random number generator, which presumably returns essentially uncorrelated floating-point numbers. This introduces randomness into the CFO algorithm. However, the CFO is a deterministic algorithm and it does not require randomness in any of its calculations. Formato (2010) suggested the following equations to restore deviated probes:

$$x'_{j,i} = max\left[x_i^{(L)} + \phi \times \left(x_{j,i}^{(old)} - x_i^{(L)}\right), x_i^{(L)}\right], \quad i = 1,2,\ldots,N \tag{14.17}$$

$$x'_{j,i} = min\left[x_i^{(U)} - \phi \times \left(x_i^{(U)} - x_{j,i}^{(old)}\right), x_i^{(U)}\right], \quad i = 1,2,\ldots,N \tag{14.18}$$

in which $x'_{j,i}$ = new value of the ith decision variable of the jth solution, j = the deviated solution, and ϕ = probe repositioning factor, which is determined by the user and is between zero and one.

14.7 Termination Criteria

The termination criterion prescribes when to terminate the algorithm. Selecting a good termination criterion has an important role on the correct convergence of the algorithm. The number of algorithmic iterations, the amount of improvement of the solution between consecutive iterations, and the run time are common termination criteria for the CFO.

14.8 User-Defined Parameters of the CFO

The size of the population (M); the value of the gravitational constants (α, β, G), ψ, and ϕ; and the termination criteria are user-defined parameters of the CFO. The initial acceleration of a probe is usually set equal to zero.

A good choice of the parameters depends on the decision space of a particular problem, and usually the optimal parameter setting for one problem is of limited utility for other problems. Consequently, determining a good set of parameters often requires performing a large number of numerical experiments. Practice and experience with specific types of problems is valuable in this respect. A reasonable method for finding appropriate values for the parameters is performing sensitivity analysis. This is accomplished by choosing combinations of parameters and running the algorithm several times for each combination. A comparison of the results from different runs helps in determining appropriate parameter values.

14.9 Pseudocode of the CFO

```
Begin
  Input parameters of the algorithm and initial data
  Generate M initial possible solutions
    deterministically
  Initialize the first acceleration of all solutions
  While (the termination criteria are not satisfied)
     Evaluate fitness value of solutions
     For j = 1 to M
         Move probe j to new position
         If the new position is outside of decision
           space
             Modify the position of solution j
         End if
     Next j
     For j = 1 to M
         Evaluate new acceleration of probe j
     Next j
  End while
  Report the population
End
```

14.10 Conclusion

This chapter reviewed the analogy between Newton's gravitational low and the CFO and explained the fundamentals of the CFO algorithm. A pseudocode of the algorithm closed the chapter's theory.

References

Formato, R. A. (2007). "Central force optimization: A new metaheuristic with applications in applied electromagnetics." Progress in Electromagnetics Research, 77, 425–491.

Formato, R. A. (2010). “Parameter-free deterministic global search with simplified central force optimization.” In: Huang, D.-S., Zhao, Z., Bevilacqua, V., and Figueroa, J. C. (Eds.), Advanced intelligent computing theories and applications, Lecture notes in computer science, Vol. 6215, Springer, Berlin, Heidelberg, 309–318.
Formato, R. A. (2011). “Central force optimization with variable initial probes and adaptive decision space.” Applied Mathematics and Computation, 217(21), 8866–8872.
Formato, R. A. (2012). “Improving bandwidth of Yagi-Uda arrays.” Wireless Engineering and Technology, 3(1), 18–24.
Haghighi, A. and Ramos, H. M. (2012). “Detection of leakage freshwater and friction factor calibration in drinking networks using central force optimization.” Water Resources Management, 26(8), 2347–2363.
Liu, Y. and Tian, P. (2015). “A multi-start central force optimization for global optimization.” Applied Soft Computing, 27(C), 92–98.
Mahmoud, K. R. (2011). “Central force optimization: Nelder-mead hybrid algorithm for rectangular microstrip antenna design.” Electromagnetics, 31(8), 578–592.

15

Biogeography-Based Optimization

Summary

This chapter describes the biogeography-based optimization (BBO), which is inspired by the science of biogeography and a meta-heuristic optimization algorithm. This chapter presents a brief literature review of the BBO and its applications and reviews the discipline of biogeography and its analogy to BBO. The BBO algorithm is described in detail, and a pseudocode of the BBO algorithm closes the chapter.

15.1 Introduction

Simon (2008) introduced the biogeography-based optimization (BBO) algorithm utilizing biogeographic concepts. Savsani et al. (2014) studied the effect of hybridizing the BBO technique with artificial immune algorithm (AIA) and the ant colony optimization (ACO). Niu et al. (2014) proposed a BBO algorithm with mutation strategies (BBO-M), which employs mutation motivated by the differential evolution (DE) algorithm and chaos theory for improving the global searching capability of the algorithm. Gupta et al. (2015) implemented the BBO for optimal component sizing of off-grid small autonomous hybrid power systems (SAHPS) by minimizing the cost of energy. Yang (2015) proposed a modified biogeography-based optimization (MBBO) algorithm to solve a flexible job shop scheduling problem (FJSSP). Tamjidy et al. (2015) used the BBO to deal with hole-making process problem. Bozorg-Haddad et al. (2015) used the BBO to optimal operation of single- and multi-reservoir systems.

Meta-Heuristic and Evolutionary Algorithms for Engineering Optimization,
First Edition. Omid Bozorg-Haddad, Mohammad Solgi, and Hugo A. Loáiciga.

15.2 Mapping Biogeography-Based Optimization (BBO) to Biogeography Concepts

Biogeography is the study of the geographical distribution of living organisms. Mathematical biogeographic models attempt to explain how species migrate between habitats, their appearance, adaptation, evolution, and extinction. The habitats that are more suitable places for species settlement have a relatively high habitat suitability index (HSI) that depends on factors such as vegetative cover, precipitation, area, temperature, and so on. Variables that determine the quality of habitat are known as suitability index variables (SIVs). SIVs are independent variables and the HSI is variable dependent on SIVs. Habitats with large values of HSI accommodate more species, and, conversely, a low HSI habitat supports fewer species. Habitats with a high HSI have many species that emigrate to nearby habitats, simply by virtue of the large number of species that they host and at the same time exhibit low species immigration rate because they already house many species. There is a stronger tendency for species to emigrate from a habitat as its number of species increases to find a new habitat with lower population density. Habitats with low population density may attract immigration provided that the habitat has adequate life-supporting characteristics. Habitats with a low HIS may have a high species immigration rate. This immigration of new species to low HSI habitats may raise the HSI of the habitat, because the suitability of a habitat is proportional to its biological diversity.

Figure 15.1 illustrates the effect that the number of species has on the immigration rate (λ) and emigration rate (μ). According to Figure 15.1 the maximum rate of immigration to the habitat occurs when there are no species in it. As the number of species in the habitat increases, the rate of immigration decreases. The rate of immigration becomes nil when the number of species in the habitat

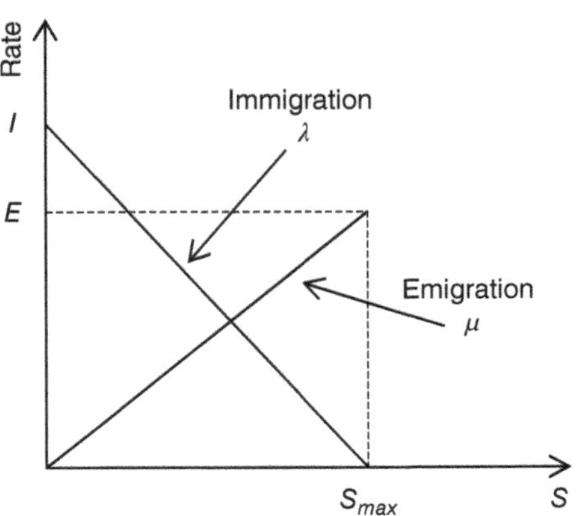

Figure 15.1 Species immigration and emigration pattern in a habitat.

Table 15.1 The characteristics of the BBO.

General algorithm (see Section 2.13)	Biogeography-based optimization
Decision variable	Suitability index variables
Solution	Habitat
Old solution	Old habitat
New solution	Modified habitat
Best solution	Habitat with max HSI (elite habitat)
Fitness function	Habitat suitability index
Initial solution	Random habitat
Selection	Migration process
Process of generating new solutions	Migration process/mutation

equals S_{max}. The rate of emigration increases as the number of species in a habitat increases, starting with zero emigration rate for an empty habitat. The maximal rates of immigration and emigration are identified by I and E, respectively. The immigration and emigration curves in Figure 15.1 are shown as straight lines but they might be nonlinear curves.

BBO designates each habitat as a solution of the optimization problem, and features of a habitat that determine its quality represent decision variables. Therefore, each SIV is a decision variable. A good solution is known as a habitat with a high HSI, while a poor solution represents a habitat with a low HSI. Also, it is assumed that high HSI solutions represent habitats with many species, whereas low HSI solutions represent habitats with few species. In nature, species travel between habitats according to the immigration and emigration rates. Therefore, solutions obtained with the BBO share their variables with each other based on their fitness values. In this manner good solutions tend to share their variables (features) with worse solutions, and poor solutions accept a lot of new variables from good solutions. In contrast, good solutions accept few variables from other solutions. Solutions that have better fitness values than others resist change more than worse solutions. This sharing of variables is intended to raise the quality of solutions. Modified solutions are new solutions. A habitat's HSI can change suddenly due to cataclysmic events affecting natural habitats such as large floods, disease, earthquakes, forest fires, and so on. The BBO simulates these events randomly in terms of mutation. Table 15.1 lists the characteristics of the BBO.

BBO begins by generating randomly a set of habitats. Each habitat is a potential solution to the given problem. The fitness value of each habitat is evaluated and mapped to the number of species, the immigration rate λ, and the emigration rate μ. Thereafter, the migration process is implemented to modify each non-elite

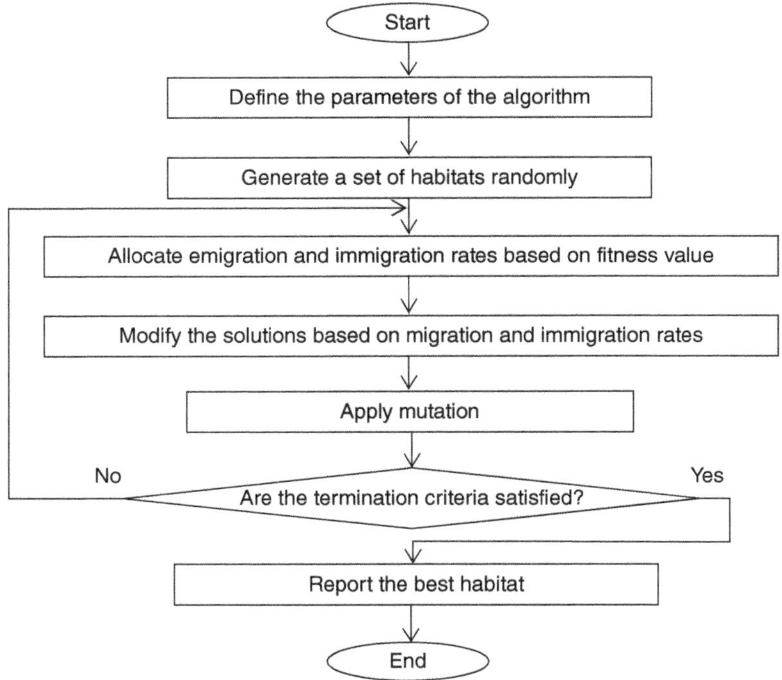

Figure 15.2 The flowchart of the BBO.

habitat, followed by its mutation. The objective function for all solutions is evaluated, and the migration and mutation process are applied to modify the habitats. This algorithm is terminated after a predefined number of iterations. The flowchart of the FA is shown in Figure 15.2.

15.3 Creating an Initial Population

BBO designates a habitat as a possible solution of the optimization problem, and each SIV of a habitat is a decision variable of the optimization problem. A habitat is represented as an array of size $1 \times N$. In an N-dimensional problem this array is written as follows:

$$Habitat = X = \left(x_1, x_2, \ldots, x_i, \ldots, x_N\right) \tag{15.1}$$

where X = a solution of optimization problem, x_i = ith decision variable of solution X, and N = number of decision variables. The decision variable values $(x_1, x_2, x_3, \ldots, x_N)$ can be represented as floating point number (real values) or as a predefined set of values for continuous and discrete problems.

The biogeography algorithm begins with the generation of a matrix of size $M \times N$ (see Section 2.6), where M and N denote the size of the population of solutions and the number of decision variables, respectively. Hence, the matrix of possible solutions that is generated randomly is written as follows (rows and columns are the number of habitats (or solutions) and the number of decision variables, respectively):

$$Population = \begin{bmatrix} X_1 \\ X_2 \\ \vdots \\ X_j \\ \vdots \\ X_M \end{bmatrix} = \begin{bmatrix} x_{1,1} & x_{1,2} & \cdots & x_{1,i} & \cdots & x_{1,N} \\ x_{2,1} & x_{2,2} & \cdots & x_{2,i} & \cdots & x_{2,N} \\ & & & \vdots & & \\ x_{j,1} & x_{j,2} & \cdots & x_{j,i} & \cdots & x_{j,N} \\ & & & \vdots & & \\ x_{M,1} & x_{M,2} & \cdots & x_{M,i} & \cdots & x_{M,N} \end{bmatrix} \tag{15.2}$$

in which $X_j = j$th solution, $x_{j,i} = i$th decision variable of the jth solution, and M = population size.

15.4 Migration Process

The HSI plays the role of a fitness value in the BBO algorithm. The greater the HSI, the more suitable the habitat is. The number of species (S) has a direct relation to the HSI for a solution (habitat) whose emigration E and immigration (I) rates are equal ($E = I$), as shown in the Figure 15.3. The HSI values can be used for evaluating the fitness of a solution. In Figure 15.3 S_1 is a solution with low HSI, while S_2 represents a high HSI solution. S_1 represents a habitat with few species, while S_2 denotes a habitat with numerous species. The λ_1 associated with S_1 is larger than the λ_2 corresponding to S_2. μ_1 for S_1 is smaller than μ_2 for S_2.

The rates of emigration (μ) and immigration (λ) are expressed in terms of the number S of species found within the habitat in the following form:

$$\mu = E \times \frac{S}{S_{max}} \tag{15.3}$$

$$\lambda = I \times \left(1 - \frac{S}{S_{max}}\right) \tag{15.4}$$

in which μ = emigration rate, E = the maximum emigration rate, λ = immigration rate, I = the maximum immigration rate, S = the number of species in the habitat, and S_{max} = the maximum number of species at which $\lambda = 0$ and $\mu = E$. Notice that $\mu + \lambda = E = I$ if $E = I$.

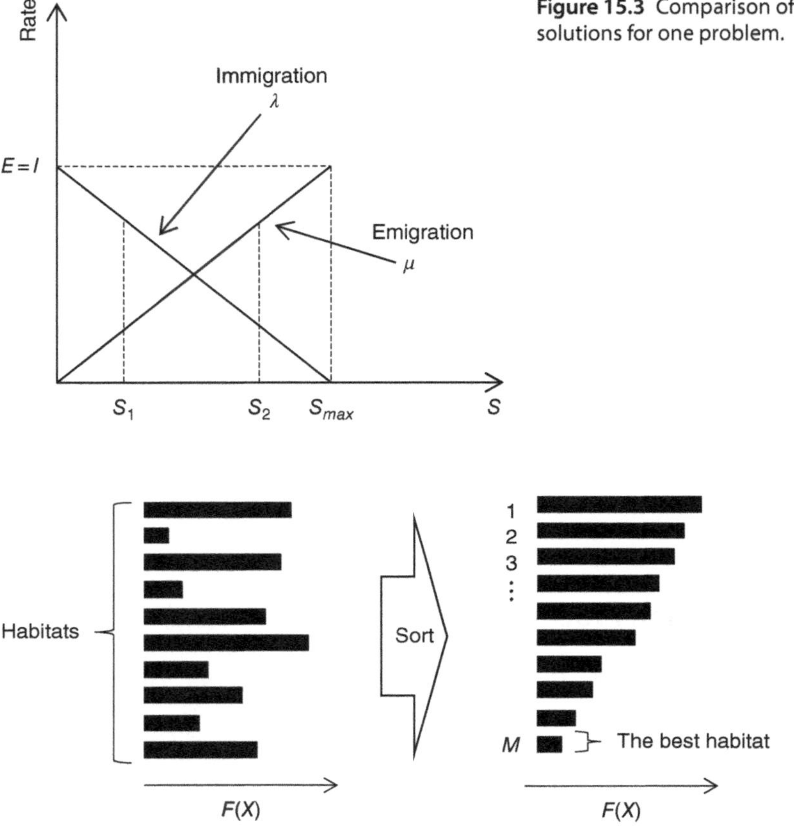

Figure 15.3 Comparison of two solutions for one problem.

Figure 15.4 Ranking habitats (solutions) according to their fitness values in a minimization problem.

The immigration process is implemented by first ranking all solutions based on their fitness values so that the best (fittest) solution is the Mth solution of the population as shown in Figure 15.4 for a minimization problem.

It is then assumed that $S_{max} = M$ and the rank of each solution, which is proportional to its fitness value is equal to S for that solution. The values μ and λ are evaluated as follows:

$$\mu_j = E \times \frac{j}{M}, \quad j = 1, 2, \ldots, M \tag{15.5}$$

$$\lambda_j = I \times \left(1 - \frac{j}{M}\right), \quad j = 1, 2, \ldots, M \tag{15.6}$$

in which μ_j = the probability of emigration of the jth solution, λ_j = the probability of immigration of the jth solution, E = the maximum probability of emigration $(0 \le E \le 1)$, I = the maximum probability of immigration $(0 \le I \le 1)$, and j = the counter of solutions that are ranked in descending order (if the problem solved is one of minimization) based on their desirability. The values of E and I are determined by the user. Lastly, for every solution j of the population, a random number in the range [0–1] ($Rand_j$) is generated. If the generated random number is less than λ_j, this means that the jth solution has experienced immigration. For any other solution r of the population, a random number in the range [0–1] ($Rand_r$) is generated. If $Rand_r$ is less than μ_r, a randomly selected decision variable of the jth solution is replaced with the corresponding decision variable of the rth solution. For example, if $X = (x_1, x_2, \ldots, x_i, \ldots, x_N)$ is selected for immigration and $X' = (x'_1, x'_2, \ldots, x'_i, \ldots, x'_N)$ is selected for emigration, and if decision variable i (x_i) is selected for replacement, then the improved solution is constructed as follows:

$$X^{(improved)} = \left(x_1, x_2, \ldots, x'_i, \ldots, x_N\right) \tag{15.7}$$

in which $X^{(improved)}$ = the improved form of solution X.

Each solution can be improved by another solution with some probability. A solution is chosen as an improvement according to its immigration rate after selecting the SIVs to be modified. The emigration rate (μ) relevant to other solutions is used to select the improvement solution. SIVs from chosen solutions are randomly replaced with the SIVs of the improvement solution. The BBO takes advantage of elitism when $\lambda = 0$ for the best habitat.

The BBO migration strategy differs from other evolutionary strategies in at least one important aspect. In evolutionary strategies global recombination such as crossover in the genetic algorithm (GA) is employed to generate new solutions. The BBO, on the other hand, implements migration to change existing solutions. Migration in the BBO, an adaptive process, is applied to modify existing habitats. On the other hand, global recombination in evolutionary strategy is a reproductive process. The BBO takes advantage of elitism in order to retain the best solutions in the population similarly to other population-based optimization algorithms. This prevents the best solutions from being destroyed by immigration.

15.5 Mutation

Events such as the spreading of infectious diseases, natural hazards, and other calamities can rapidly change the HSI of a habitat. Therefore, the condition of a habitat changes from adequate to inadequate in a manner similar to mutations in the GA. Mutation can be exerted on SIVs after migration based on a probability distribution such as the Gaussian distribution or the uniform distribution. The mutation operator replaces some of the decision variables with randomly generated values. Let $X = (x_1, x_2, \ldots, x_i, \ldots, x_N)$ be a solution of an optimization problem and assume that the ith decision variable (x_i) is selected for mutation. The mutated solution $X = (x_1, x_2, \ldots, x_i', \ldots, x_N)$ obtained by producing x_i' is calculated as follows:

$$x_i' = Rnd\left(x_i^{(L)}, x_i^{(U)}\right) \tag{15.8}$$

in which x_i' = the new value of x_i that is selected for mutation, $x_i^{(U)}$ = the upper bound of the ith decision variable, $x_i^{(L)}$ = the lower bound of the ith decision variable, and $Rnd(a,b)$ = a random value in the range $[a,b]$.

A probability of mutation (P_c) is applied to solutions to implement mutation. The elite solution has a probability of mutation equal to zero. A random number is generated for every solution. If the generated random number is less than P_c, that solution experiences mutation.

15.6 Termination Criteria

The termination criterion prescribes when to terminate the algorithm. Selecting a good termination criterion has an important role on the correct convergence of the algorithm. The number of algorithmic iterations, the amount of improvement of the solution between consecutive iterations, and the run time are common termination criteria for the CFO.

15.7 User-Defined Parameters of the BBO

The population size (M), the maximal probability of immigration (I) and emigration (E), the probability of mutation (P_c), and the termination criterion are user-defined parameters of the BBO. A good choice of the parameters depends on the decision space of a particular problem, and commonly the optimal set of parameter for one problem is of limited utility for other problems. Practice and experience with specific types of problems is valuable in this respect. A reasonable method for finding appropriate values for the parameters is performing sensitivity analysis. This is accomplished by choosing combinations of parameters and running the algorithm several times for each combination. A comparison of the results from different runs helps in determining appropriate parameter values.

15.8 Pseudocode of the BBO

```
Begin
  Input the parameters of the algorithm and initial
    data
  Generate M initial possible solutions randomly
  While (the termination criteria are not satisfied)
     Evaluate fitness value of solutions
     Rank solutions based on their fitness values so
       that Mth solution is the fittest solution
     Determine the immigration (λ) and emigration (μ)
       probabilities based on the ranks of solutions
     For j = 1 to (M - 1)
         Select habitat j with probability λj
         If habitat j is selected
             For r = 1 to M
                 Select habitat r with probability μr
                 If habitat r is selected
                     Select xj,i randomly
                     Put xj,i = xr,i
                 End if
             Next r
         End if
     Next j
     For j = 1 to (M - 1)
         Select habitat j with probability Pc
         If habitat j is selected
             Put xj,i equal to a random value
         End if
     Next j
  End while
  Report the best solution
End
```

15.9 Conclusion

This chapter described the BBO algorithm. First, a brief literature review about the BBO and its application was presented. The principles of biogeography were described and mapped to the BBO algorithm, which was described in detail. Lastly, a pseudocode of the BBO closed the chapter.

References

Bozorg-Haddad, O., Hosseini-Moghari, S., and Loáiciga, H. A. (2015). "Biogeography-based optimization algorithm for optimal operation of reservoir systems." Journal of Water Resources Planning and Management, 142(1), 04015034.

Gupta, R. A., Kumar, R., and Bansal, A. K. (2015). "BBO-based small autonomous hybrid power system optimization incorporating wind speed and solar radiation forecasting." Renewable and Sustainable Energy Reviews, 41, 1366–1375.

Niu, Q., Zhang, L., and Li, K. (2014). "A biogeography-based optimization algorithm with mutation strategies for model parameter estimation of solar and fuel cells." Energy Conversion and Management, 86, 1173–1185.

Savsani, P., Jhala, R. L., and Savsani, V. (2014). "Effect of hybridizing biogeography-based optimization (BBO) technique with artificial immune algorithm (AIA) and ant colony optimization (ACO)." Applied Soft Computing, 21, 542–553.

Simon, D. (2008). "Biogeography-based optimization." IEEE Transactions on Evolutionary Computation, 12(6), 702–713.

Tamjidy, M., Paslar, S., Baharuding, B. T. H. T., Hong, T. S., and Aiffin, M. K. A. (2015). "Biogeography based optimization (BBO) algorithm to minimize non-productive time during hole-making process." International Journal of Production Research, 53(6), 1880–1894.

Yang, Y. (2015). "A modified biogeography-based optimization for the flexible job shop scheduling problem." Mathematical Problems in Engineering, 2015, 184643.

16

Firefly Algorithm

Summary

This chapter describes the firefly algorithm (FA), which is inspired by the flashing powers of fireflies. It is a meta-heuristic optimization algorithm. This chapter presents in sequence a brief literature review of the FA and its applications, the characteristics of fireflies and their mapping to the FA, a detailed description of the FA, and a pseudocode of the FA.

16.1 Introduction

Yang (2008) introduced the firefly algorithm (FA) and applied it to solve several optimization test problems whose results compared favorably with the genetic algorithm (GA) and particle swarm optimization (PSO) (Yang, 2009). Yang (2010) merged the Levy flight (LF) approach searching with the FA and solved several optimization test problems by applying the proposed hybrid algorithm. The results indicated that the success rate of the Levy flight FA (LFA) was better than that of the standard FA. Yang (2011) applied chaos theory for auto-tuning of the parameters of the algorithm. The results of the cited study compared favorably with those of the standard FA for the well-known problem of the welded beam. Yan et al. (2012) developed an adaptive FA (AFA) to upgrade the FA's capability in solving large dimensional. The latter authors showed that the AFA performed better with several test problems than the standard FA, differential evolution (DE), and PSO. Many studies have been devoted to improving the searching accuracy of the FA and have shown its better convergence rate than other algorithms. The advantage of the FA from the standpoint of speed of convergence has led to its adoption in solving complex and nonlinear problems in different scientific fields. In this context, Afnizanfaizal et al. (2012) introduced a new hybrid FA named

Meta-Heuristic and Evolutionary Algorithms for Engineering Optimization,
First Edition. Omid Bozorg-Haddad, Mohammad Solgi, and Hugo A. Loáiciga.

hybrid evolutionary FA (HEFA) to improve the searching accuracy of the original FA. This approach was a combination of the FA and the DE algorithm with the goal of estimating the parameters of a nonlinear and complex biological model of large dimensionality. The results showed that HEFA has an improved searching accuracy compared with the GA, PSO, and evolutionary programming (EP). Santos et al. (2013) calculated the amount of precipitation of a region in South America. They computed the precipitation using six different methods. In each of these methods, different effective parameters were used to calculate the precipitation. The FA was applied to find the optimal weights for the various methods. In a comprehensive review of the FA, Fister et al. (2013) concluded that the FA's solving efficiency is explained by its capacity to solve multimodal, nonlinear, optimization problems. Garousi-Nejad et al. (2016b) applied the FA to reservoir operation and demonstrates the superiority of this algorithm against the GA. Garousi-Nejad et al. (2016a) presented a modified FA for solving multi-reservoir operation in continuous and discrete domains.

16.2 Mapping the Firefly Algorithm (FA) to the Flashing Characteristics of Fireflies

There are about 2000 firefly species, most of which produce short and rhythmic flashes. The flashing light of fireflies is an interesting sight in the summer sky of tropical and temperate areas. Usually a particular species exhibits a unique flashing pattern. The flashing light is generated by bioluminescence. It is believed that two fundamental functions of such flashes are to attract mating partners (communication) and to attract potential prey. In addition, flashing may also serve as a protective warning mechanism. Several factors including the rhythmic flash, the rate of flashing, and the duration of flashing form part of the signal system that brings both sexes together. Females respond to a male's unique pattern of flashing in the same species, while in some species such as *Photuris*, female fireflies can mimic the mating flashing pattern of other species to lure and eat the male fireflies who may mistake the flashes as a potential suitable mate. In summary, fireflies flash their stored energy as a light to mate, hunt, or evade predators. Fireflies produce attractiveness by shining light.

It is known that the light intensity at a particular distance from the light source follows the inverse square law, whereby the light intensity decreases with increasing distance between a viewer and the source of the light. Furthermore, the air absorbs light that becomes weaker as the distance increases. Fireflies are thus visible only over a restricted distance, usually several hundred meters in the dark, which is usually sufficient for fireflies to communicate.

The FA assumes that the flashing light can be formulated in such a way that it is associated with the objective function of the optimization problem. The FA is based on three idealized rules:

1) All fireflies are unisex so their attractiveness depends on the amount of light flashed by them regardless of their sex.
2) The attractiveness of fireflies is proportional to their brightness. Thus, for any two flashing fireflies, the firefly that flashes less will move toward the firefly that flashes more. The attractiveness and the brightness of fireflies decrease as the distance between fireflies increases. Thus, the movement of fireflies continues in this manner until there is no brighter firefly in a group. Once this happens the fireflies move randomly.
3) The brightness of a firefly is determined by a fitness function.

The FA designates a firefly as a solution whose location in any N-dimensional is a decision variable. In nature each firefly moves toward other fireflies according to their attractiveness. For simplicity, it is assumed that the attractiveness of a firefly is determined by its brightness, which in turn is associated with the fitness function. The FA dictates that if the fitness value of a firefly is larger than that of another firefly, the firefly with less brightness (fitness value) moves toward the firefly with more brightness. The movement of the firefly is based on the light intensity of the other firefly, which is influenced by the distance between the fireflies. New positions occupied by the fireflies are new solutions. Table 16.1 lists the characteristics of the FA, and the flowchart of the FA is shown in Figure 16.1.

Table 16.1 The characteristics of the FA.

General algorithm (see Section 2.13)	Firefly algorithm
Decision variable	Position of firefly in each dimension
Solution	Firefly (position)
Old solution	Old position of firefly
New solution	New position of firefly
Best solution	–
Fitness function	Brightness
Initial solution	Random firefly
Selection	–
Process of generating new solutions	Movement of firefly

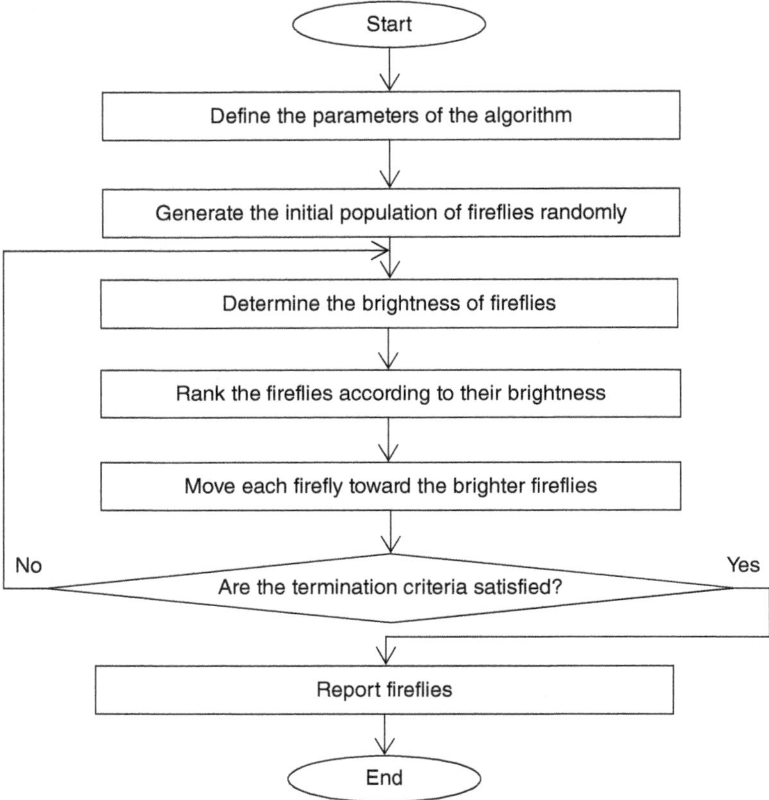

Figure 16.1 The flowchart of the FA.

16.3 Creating an Initial Population

Each possible solution of the optimization problem is called a firefly in the FA. In an N-dimensional optimization problem, a firefly's location represents a decision variable of the optimization problem. A firefly is denoted by an array of size $1 \times N$ that represents the firefly's location. This array is defined as follows:

$$Firefly = X = \left(x_1, x_2, \ldots, x_i, \ldots, x_N\right) \tag{16.1}$$

where X = a solution of optimization problem, x_i = ith decision variable of solution X, and N = number of decision variables. The decision variable values ($x_1, x_2, x_3, \ldots, x_N$) can be represented as floating point number (real values) or as a predefined set for continuous and discrete problems, respectively.

The FA algorithm starts by randomly generating a matrix of size $M \times N$ (see Section 2.6), where M and N denote the size of the population and the number of decision variables, respectively. Hence, the matrix of solutions that is generated randomly is given as follows (rows and columns are the number of fireflies and the number of decision variables, respectively):

$$Population = \begin{bmatrix} X_1 \\ X_2 \\ \vdots \\ X_j \\ \vdots \\ X_M \end{bmatrix} = \begin{bmatrix} x_{1,1} & x_{1,2} & \cdots & x_{1,i} & \cdots & x_{1,N} \\ x_{2,1} & x_{2,2} & \cdots & x_{2,i} & \cdots & x_{2,N} \\ & & & \vdots & & \\ x_{j,1} & x_{j,2} & \cdots & x_{j,i} & \cdots & x_{j,N} \\ & & & \vdots & & \\ x_{M,1} & x_{M,2} & \cdots & x_{M,i} & \cdots & x_{M,N} \end{bmatrix} \tag{16.2}$$

in which X_j = jth solution, $x_{j,i}$ = ith decision variable of the jth solution, and M = population size.

16.4 Attractiveness

The attractiveness is the brightness of the light emitted by fireflies, which varies with the squared distance between them. In addition, light intensity decreases with the distance from the source. According to Yang's (2009) assumptions, the attractiveness of a firefly at a distance d is calculated as follows:

$$\beta(d) = \beta_0 \times e^{-\gamma \times d^m}, \quad m > 1 \tag{16.3}$$

in which $\beta(d)$ = firefly's attractiveness at distance d from the firefly, β_0 = the attractiveness at a distance $d = 0$, γ = light absorption coefficient, d = the distance between any two fireflies, and m = exponent. Yang (2009) proposed the value of $m = 2$.

16.5 Distance and Movement

The distance between the kth and jth fireflies that are located at X_k and X_j positions, respectively, is computed as follows:

$$d_{j,k} = \left\| X_j - X_k \right\| = \sqrt{\sum_{i=1}^{N} \left(x_{j,i} - x_{k,i} \right)^2} \tag{16.4}$$

in which $d_{j,k}$ = Cartesian distance between the jth and kth fireflies, $\| \;\|$ = the magnitude of the distance vector between the jth and kth fireflies in space, $x_{j,i}$ = ith dimension of the spatial coordinate of the jth firefly's position (ith decision variable of the jth solution), N = number of dimensions (decision variables), and $x_{k,i}$ = ith dimension of the spatial coordinate of the kth firefly's position. It is worth mentioning that $d_{j,k}$ defined in Equation (16.4) is not limited to the Euclidean distance. In fact, any measure that can effectively characterize the quantities of interests in the optimization problems can be used as the distance depending on the type of the problem at hand (Yang, 2013).

For any two flashing fireflies, the firefly that flashes less intensely (fitness value) moves toward the firefly that flashes more intensely (fitness value). The longer the distance between fireflies, the lower their mutual attractiveness is. The movement of fireflies continues guided by these two rules until there is not a brighter firefly in a group. At that time fireflies move randomly. Rule 3 states that the brightness of a firefly is determined by a fitness function.

If a pair of fireflies k and j is to be considered so that firefly j is better than firefly k in terms of brightness (fitness value), then firefly k is attracted by firefly j and will move toward the position of firefly j. As the result of this movement, firefly k would move to a new position that is computed as follows:

$$X_k^{(new)} = X_k + \beta_0 e^{-\gamma d_{j,k}^2}(X_j - X_k) + \alpha(Rand - 0.5) \tag{16.5}$$

in which $X_k^{(new)}$ and X_k = new position and current position of firefly k that has less brightness (solution with worse fitness value), respectively, X_j = position of firefly j that has more brightness (solution with better fitness value), α = a randomized parameter, and *Rand* is a random value in the range [0,1]. The second and third terms of Equation (16.5) correspond to the attraction and randomization, respectively. β_0, γ, and α are parameters of the algorithm.

16.6 Termination Criteria

Termination criterion determines when to terminate the algorithm. Selecting a good termination criterion has an important role to correct convergence of the algorithm. The number of iterations, the amount of improvement of solutions between consecutive iterations, and the run time are common termination criteria for the FA.

16.7 User-Defined Parameters of the FA

The size of population (M), the initial attractiveness (β_0), the light absorption coefficient (γ), the randomized parameter (α), and the termination criteria are user-defined parameters of the FA. Yang (2009) pointed out that for most implementations, the value of β_0 equals 1. Moreover, according to Yang (2009), the range of values of α is [0,1]. Even though Yang (2013) pointed out that it is better to use a time-dependent α so that randomness decreases gradually as the iterations proceed. Also, γ is a light absorption coefficient that takes values in the range $[0,\infty)$, in theory. When $\gamma = 0$ the attractiveness is constant. In other words, the light intensity does not decrease. Therefore, a flashing firefly can be seen anywhere in the domain. In contrast, $\gamma = \infty$ means that the attractiveness is almost zero in the view of other fireflies. In practice γ is usually in the range [0.1,10] (Yang, 2009). It is worth mentioning that the value of these parameters is the key in determining the convergence speed and the overall capability of the algorithm. Thus, a sensitivity analysis of these parameters is of vital importance.

16.8 Pseudocode of the FA

```
Begin
  Input the parameters of the algorithm and initial data
  Generate M initial possible solutions randomly
  While (the termination criteria are not satisfied)
     Determine fitness value of all solutions
     Sort all solutions according to their fitness
       values
     For k = 1 to M
         For j = 1 to M
             If F(X_j) is better than F(X_k)
                  Move the solution k toward the solution j
             End if
         Next j
     Next k
  End while
  Report all solutions
End
```

16.9 Conclusion

This chapter described the FA, presented a literature review of the FA and its application, mapped the characteristics of fireflies into the FA, described the FA in detail, and closed with a pseudocode of the FA.

References

Afnizanfaizal, A., Safaai, D., Mohd Saberi, M., and Siti Zaiton, M. H. (2012). “A new hybrid firefly algorithm for complex and nonlinear problem.” In: Omatu, S., Santana, J. F. D. P., González, S. R., Molina, J. M., Bernardos, A. M., and Rodríguez, J. M. C. (Eds.), Distributed computing and artificial intelligence, Advances in intelligent and soft computing, Vol. 151, Springer, Berlin, Heidelberg, 673–680.

Fister, I., Fister, I., Jr., Yang, X. S., and Brest, J. (2013). "A comprehensive review of firefly algorithms." Swarm and Evolutionary Computation, 13(1), 34–46.

Garousi-Nejad, I., Bozorg-Haddad, O., and Loáiciga, H. A. (2016a). "Modified firefly algorithm for solving multireservoir operation in continuous and discrete domains." Journal of Water Resources Planning and Management, 142(9), 04016029.

Garousi-Nejad, I., Bozorg-Haddad, O., Loáiciga, H. A., and Mariño, M. A. (2016b). "Application of the firefly algorithm to optimal operation of reservoirs with the purpose of irrigation supply and hydropower production." Journal of Irrigation and Drainage Engineering, 142(10), 04016041.

Santos, A. F., Campos Velho, H. F., Luz, E. F., Freitas, S. R., Grell, G., and Gan, M. A. (2013). "Firefly optimization to determine the precipitation field on South America." Inverse Problem in Science and Engineering, 21(3), 451–466.

Yan, X., Zhu, Y., Wu, J., and Chen, H. (2012). "An improved firefly algorithm with adaptive strategies." Advanced Science Letters, 16(1), 249–254.

Yang, X. S. (2008). "Nature-inspired meta-heuristic algorithms." Luniver Press, Beckington.

Yang, X. S. (2009). "Firefly algorithm for multimodal optimization." Stochastic Algorithms: Foundations and Applications, 5792(2), 169–178.

Yang, X. S. (2010). "Firefly algorithm, Lévy flights and global optimization." In: Bramer, M., Ellis, R., and Petridis, M. (Eds.), Research and development in intelligent systems XXVI, Springer, London, 209–218.

Yang, X. S. (2011). "Chaos-enhanced firefly algorithm with automatic parameter tuning." Journal of Swarm Intelligence Research, 2(4), 1–11.

Yang, X. S. (2013). "Multiobjective firefly algorithm for continuous optimization." Engineering with Computers, 29(2), 175–184.

17

Gravity Search Algorithm

Summary

This chapter describes the gravity search algorithm (GSA), an evolutionary optimization algorithm based on the law of gravity and mass interactions. It designates a particle as a solution of an optimization problem. Particles exhibit simple behavior, and they follow intelligent pathways toward the near-optimal solution. This chapter presents a literature review of the GSA and its applications, explains the GSA's analogy to the law of gravity and the GSA in detail, and closes with a pseudocode of the GSA.

17.1 Introduction

Rashedi et al. (2009) introduced the gravity search algorithm (GSA) based on the law of gravity and mass interactions and compared it with the particle swarm optimization (PSO) and central force optimization (CFO) with well-known benchmark functions. Their results established the excellent performance of the GSA in solving various nonlinear functions. Ghalambaz et al. (2011) presented a hybrid neural network and gravitational search algorithm (HNGSA) method to solve the well-known Wessinger's equation. Their results showed that HNGSA produced a closer approximation to the analytic solution than other numerical methods and that it could easily be extended to solve a wide range of problems. Jadidi et al. (2013) proposed a flow-based anomaly detection system and used a multilayer perceptron (MLP) neural network with one hidden layer for solving it. The latter authors optimized the interconnection weights of an MLP network with the GSA, and the proposed GSA-based flow anomaly detection system (GFADS) was trained with a flow-based data set. Chen et al. (2014) proposed an improved gravitational search algorithm (IGSA) and solved the identification problem for a water turbine regulation system (WTRS) under load and no-load running conditions.

Meta-Heuristic and Evolutionary Algorithms for Engineering Optimization,
First Edition. Omid Bozorg-Haddad, Mohammad Solgi, and Hugo A. Loáiciga.

17.2 Mapping the Gravity Search Algorithm (GSA) to the Law of Gravity

Every particle in the universe attracts every other particle because of gravity. Gravitation is the tendency of masses to accelerate toward each other. Newton's law of gravity states that particles attract each other with a force that is directly proportional to their masses and inversely proportional to the square of the distance between them. Nature encompasses three types of masses:

1) Active gravity mass, in which the gravity force increases with increasing mass
2) Passive gravity mass, in which the gravity force does not increase with increasing mass
3) Inertial mass that expresses mass resistance to changing its position and movement

Particles attract each other with a specific force that is directly related to the masses of the particles and inversely related to the square distance between their centers of mass (see Figure 17.1):

$$Force = \gamma \times \frac{Mass_1 \times Mass_2}{d^2} \tag{17.1}$$

where $Force$ = gravity force (N), $Mass_1$ = active gravity mass (kg) of first particle, $Mass_2$ = passive mass (kg) of second particle, γ = Newton's gravitational constant $[(Nm^2)/kg^2]$, and d = distance separating the centers of masses of the two particles (m).

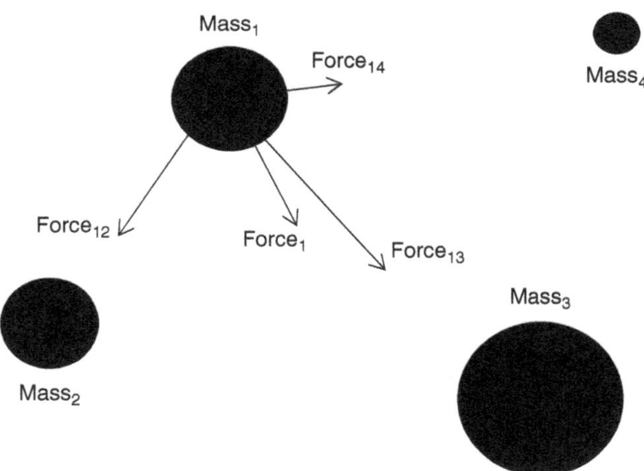

Figure 17.1 Gravity force between different particles; $Force_1$ is the resultant force on $Mass_1$.

Table 17.1 The characteristics of the GSA.

General algorithm (see Section 2.13)	Gravity search algorithm
Decision variable	Position of particle in each dimension
Solution	Position of particle
Old solution	The old position of particle
New solution	The new position of particle
Best solution	–
Fitness function	Mass of particle
Initial solution	Random particle
Selection	–
Process of generating new solutions	Movement of particle

Newton's second law states that when a force is applied to a particle, its acceleration depends only on the force and its mass:

$$A = \frac{Force}{Mass} \tag{17.2}$$

where A = particle acceleration and $Mass$ = inertial mass. Also, based on the law of motion, the summation of the velocity ratio and acceleration at time t is considered in updating the velocity at time $t+1$. The variation of the velocity or acceleration of any mass is equal to the force acting on the system divided by the inertial mass.

According to the GSA every particle in the system determines the position and state of other particles employing the law of gravity (Rashedi et al., 2009). The GSA begins by randomly choosing the positions of particles over the entire solution space. Thereafter, a mass is assigned to each particle according to its fitness value. Notice that the position of particles determines the fitness value. In the next step, the exerted force on each particle by other particles is calculated. Lastly, each particle moves to new positions based on the summation force of other particles. Table 17.1 defines the characteristics of the GSA, and the steps of the GSA are depicted in Figure 17.2.

17.3 Creating an Initial Population

Each possible solution of the optimization problem is called a particle by the GSA. In an N-dimensional optimization problem, a particle is an array of size $1 \times N$. This array is defined as follows:

$$Particle = X = \left(x_1, x_2, \ldots, x_i, \ldots, x_N\right) \tag{17.3}$$

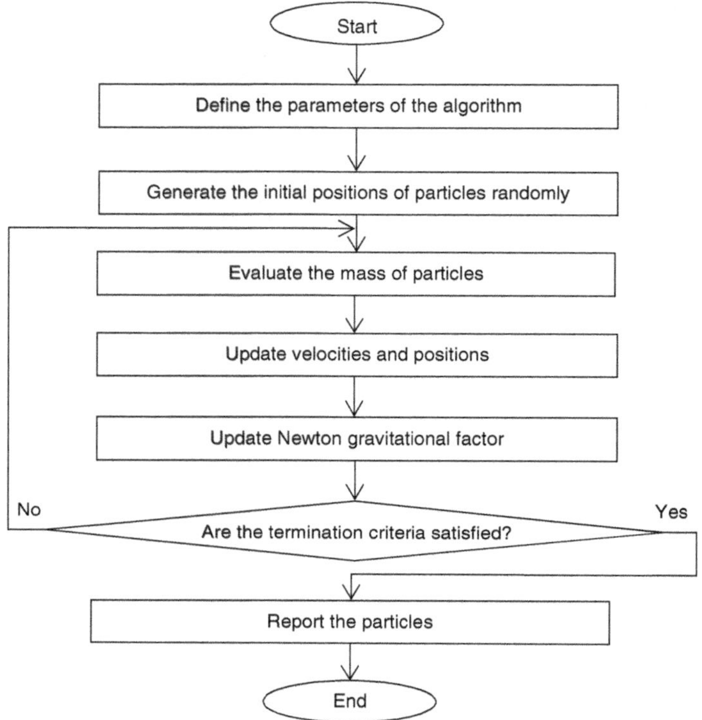

Figure 17.2 The flowchart of the GSA.

where X = a solution of optimization problem, x_i = ith decision variable of solution X, and N = number of decision variables. Each of the decision variable values $(x_1, x_2, x_3, \ldots, x_N)$ can be represented as floating point number (real values) or as a predefined set for continuous and discrete problems, respectively.

The GSA algorithm starts with the random generation (see Section 2.6) of a matrix of size $M \times N$, where M and N are the size of population and the number of decision variables, respectively. Hence, the matrix of solutions that is generated randomly is given as follows (rows and columns are the number of particles and the number of decision variables, respectively):

$$Population = \begin{bmatrix} X_1 \\ X_2 \\ \vdots \\ X_j \\ \vdots \\ X_M \end{bmatrix} = \begin{bmatrix} x_{1,1} & x_{1,2} & \cdots & x_{1,i} & \cdots & x_{1,N} \\ x_{2,1} & x_{2,2} & \cdots & x_{2,i} & \cdots & x_{2,N} \\ & & & \vdots & & \\ x_{j,1} & x_{j,2} & \cdots & x_{j,i} & \cdots & x_{j,N} \\ & & & \vdots & & \\ x_{M,1} & x_{M,2} & \cdots & x_{M,i} & \cdots & x_{M,N} \end{bmatrix} \tag{17.4}$$

in which $X_j = j$th solution, $x_{j,i} = i$th decision variable of the jth solution, and M = population size.

17.4 Evaluation of Particle Masses

The fitness value is calculated for each particle. The best and the worst values of the fitness value are called *Best* and *Worst*, respectively, which are determined as follows (under minimization):

$$Worst = \underset{j=1}{\overset{M}{Max}}\left[F\left(X_j\right)\right] \tag{17.5}$$

$$Best = \underset{j=1}{\overset{M}{Min}}\left[F\left(X_j\right)\right] \tag{17.6}$$

where $F(X_j)$ = the fitness value of the jth solution and *Worst* and *Best* = the fitness value of the worst and best solution, respectively. Then, the relative normalized fitness value is calculated as follows:

$$\psi(X_j) = \frac{F(X_j) - Worst}{Best - Worst}, \quad j = 1,2,\ldots,M \tag{17.7}$$

where $\psi(X_j)$ = the normalized fitness value of solution j. The mass of each particle is calculated based on the normalized fitness value as follows:

$$Mass(X_j) = \frac{\psi(X_j)}{\sum_{j=1}^{M}\psi(X_j)}, \quad j = 1,2,\ldots,M \tag{17.8}$$

in which $Mass(X_j)$ = mass of particle (solution) jth. It is clear that the value of *Mass* increases with increasing the difference between the fitness value of solution j and *Worst*.

17.5 Updating Velocities and Positions

Equation (17.9) is employed to calculate the force acting on the jth particle exerted by the rth particle:

$$Force_{j,r,i} = \gamma \times \frac{Mass(X_j) \times Mass(X_r)}{d_{j,r} + \varepsilon}\left(x_{j,i} - x_{r,i}\right)$$
$$j = 1,2,\ldots,M, \quad r = 1,2,\ldots,M, \quad i = 1,2,\ldots,N \tag{17.9}$$

where $Force_{j,r,i}$ = the force action on the jth particle by the rth particle in the ith dimension, ε = small positive constant, and $d_{j,r}$ = the Euclidean distance between the jth mass and rth mass that is calculated as follows:

$$d_{j,r} = \sqrt{\sum_{i=1}^{N}\left(x_{j,i} - x_{r,i}\right)^2}, \quad j = 1,2,\ldots,M, \quad r = 1,2,\ldots,M \tag{17.10}$$

The GSA algorithm is randomized by assuming that the total force acting on the jth particle in the ith dimension is a randomly weighted sum of the ith components of the forces exerted by other particles. The acceleration of each mass in the ith dimension is calculated based on the second law of motion as follows:

$$a_{j,i} = \frac{\sum_{r=1}^{M}\left(Rand \times Force_{j,r,i}\right)}{Mass\left(X_j\right)}, \quad j = 1,2,\ldots,M, \quad i = 1,2,\ldots,N \tag{17.11}$$

where $a_{j,i}$ = acceleration of the jth particle (solution) in the ith dimension and $rand$ = a random number with uniform distribution in the interval [0,1] that introduces random properties to the GSA.

The velocity is calculated as follows:

$$v_{j,i}^{(new)} = Rand \times v_{j,i} + a_{j,i}, \quad j = 1,2,\ldots,M, \quad i = 1,2,\ldots,N \tag{17.12}$$

where $v_{j,i}^{(new)}$ = new velocity of the jth particle (solution) in the ith dimension, $v_{j,i}$ = previous velocity of the jth particle (solution), and $Rand$ = a uniform random variable in the interval [0,1].

The new position of the jth solution is given by

$$x'_{j,i} = x_{j,i} + v_{j,i}^{(new)}, \quad j = 1,2,\ldots,M, \quad i = 1,2,\ldots,N \tag{17.13}$$

$$X_j^{(new)} = \left(x'_{j,1}, x'_{j,2}, \ldots, x'_{j,i}, \ldots, x'_{j,N}\right) \tag{17.14}$$

where $x'_{j,i}$ = new value of the ith decision variable of the jth solution, $x_{j,i}$ = ith decision variable of the jth solution, and $X_j^{(new)}$ = new position of the jth particle (new solution).

17.6 Updating Newton's Gravitational Factor

The factor γ is a parameter that controls the searching capacity and the GSA's efficiency. The searching capacity of the optimization algorithm increases whenever γ increases. On the other hand, the convergence efficiency of the search algorithm increases when γ decreases. For these reasons, it is recommendable to use a value of γ that is set initially high and decreases with increasing time (Rashedi et al., 2009). A suitable formula for γ is the following:

$$\gamma^{(t)} = \gamma_0 \times e^{-\frac{C \times t}{T}}, \quad t = 1,2,\ldots,T \tag{17.15}$$

where $\gamma^{(t)}$ = Newton gravitational constant in iteration t, γ_0 and C = controlling coefficients of the GSA, t = current iteration, and T = lifetime of the system (total number of iterations). In Equation (17.15), $\gamma^{(t)}$ is initialized at the beginning of the optimization and is reduced with advancing time to control the search accuracy.

17.7 Termination Criteria

The termination criterion determines when to terminate the algorithm. Selecting a good termination criterion has an important role on the correct convergence of the algorithm. The number of iterations, the amount of improvement of the solution between consecutive iterations, and the run time are common termination criteria for the GSA.

17.8 User-Defined Parameters of the GSA

The population size (M), the initial Newton gravitational constant (γ_0), C, and the termination criteria are user-defined parameters of the GSA. A good choice of the parameters depends on the decision space of a particular problem, and usually the optimal parameter setting for one problem is of limited utility for other problems. Determining a good set of parameter often requires performing computational experiments. A reasonable method for finding appropriate values for the parameters is performing sensitivity analysis, whereby combinations of parameters are tested and the algorithm is run several times for each combination to account for the random nature of the solution algorithm. In this manner the analyst obtains a distribution of solutions and associated objective function values for each combination of parameters. A comparison of the results from all the combination of parameters provides guidance on a proper choice of the algorithmic parameters.

17.9 Pseudocode of the GSA

```
Begin
  Input the parameters of the algorithm and initial data
  Generate M initial possible solutions randomly
  While (the termination criteria are not satisfied)
```

```
    Determine the best and worst solution according
      to the fitness value
    For j = 1 to M
        Evaluate inertial mass of solution j
    Next j
    For j = 1 to M
        For r = 1 to M
            Evaluate Euclidian distance between two
              solution j and r
            For i = 1 to N
                Calculate the force action on solution
                  j from solution r in dimension i
            Next i
        Next r
        Update the acceleration and velocity of
          solution j
        Move solution j to new position
    Next j
    Update newton gravitational factor
  End while
  Report all solutions
End
```

17.10 Conclusion

This chapter described the GSA, presented a brief review of the GSA and its applications, described analogies between the GSA and the law of gravity, explained the GSA in detail, and introduced a pseudocode for the GSA.

References

Chen, Z., Yuan, X., Tian, H., and Ji, B. (2014). "Improved gravitational search algorithm for parameter identification of water turbine regulation system." Energy Conversion and Management, 78, 306–315.

Ghalambaz, M., Noghrehabadi, A. R., Behrang, M. A., Assareh, E., Ghanbarzadeh, A., and Hedayat, N. (2011). "A hybrid neural network and gravitational search algorithm (HNNGSA) method to solve well known Wessinger's equation." World Academy of Science Engineering and Technology, 49(51), 803–807.

Jadidi, Z., Muthukkumarasamy, V., Sithirasenan, E., and Sheikhan, M. (2013). "Flow-based anomaly detection using neural network optimized with GSA algorithm." IEEE 33rd International Conference on Distributed Computing Systems Workshops, Philadelphia, PA, July 8–11, Piscataway, NJ: Institute of Electrical and Electronics Engineers (IEEE), 76–81.

Rashedi, E., Nezamabadi-Pour, H., and Saryazdi, S. (2009). "GSA: A gravitational search algorithm." Information Sciences, 179(13), 2232–2248.

18

Bat Algorithm

Summary

This chapter describes the bat algorithm (BA) that is a relatively new meta-heuristic optimization algorithm. The basic concepts of the BA are inspired by the echolocation behavior of bats. The following sections present a literature review of the BA and its applications, a description of the analogy between the behavior of microbats and the BA, and a detailed explanation of the BA and introduce a pseudocode of the BA.

18.1 Introduction

Yang (2010) developed the bat algorithm (BA) based on the echolocation features of microbats. The continuous optimization of engineering design optimization has been extensively studied with the BA, which demonstrated that the BA can deal with highly nonlinear problems efficiently and can find the optimal solutions accurately (Yang, 2010, 2012; Yang and Gandomi, 2012). Case studies include pressure vessel design, automobile design, spring and beam design, truss systems, tower and tall building design, and others. Assessments of the BA features are found in Koffka and Ashok (2012), who compared the BA with the genetic algorithm (GA) and particle swarm optimization (PSO) in cancer research problems and provided evidence that the BA performs better than the other two algorithms. Malakooti et al. (2012) implemented the BA to solve two types of multiprocessor scheduling problems (MSP) and concluded that bat intelligence outperformed the list algorithm and the GA in the case of single-objective MSP. Reddy and Manoj (2012) applied fuzzy logic and the BA to obtain optimum capacitor placement for loss reduction in electricity distribution systems. Ramesh et al. (2013) reported a detailed study of combined economic load and emission

Meta-Heuristic and Evolutionary Algorithms for Engineering Optimization,
First Edition. Omid Bozorg-Haddad, Mohammad Solgi, and Hugo A. Loáiciga.

dispatch problems employing the BA. They compared this algorithm with the ant colony optimization (ACO) algorithm, hybrid GA, and other methods and concluded that the BA is easy to implement and much superior to the comparison algorithms in terms of accuracy and efficiency. Niknam et al. (2013) showed that the BA outperforms the GA and PSO in solving energy-generation problems. Baziar et al. (2013) compared the BA with the GA and PSO in the management of micro-grid for various types of renewable power sources and concluded that the BA has the best performance. Bozorg-Haddad et al. (2014) applied the BA to find optimal operation of water reservoir systems.

18.2 Mapping the Bat Algorithm (BA) to the Behavior of Microbats

Bats, the only winged mammals, can determine their locations while flying by sound emission and reception, which is called echolocation. Their population amounts to about 20% of all mammal species. Bat sizes range from the tiny bumblebee bat (with mass ranging from 1.5 to 2 g) to the giant bats with wingspan of about 2 m weighing about 1 kg (Altringham, 1996; Colin, 2000).

Most microbats are insectivores and use a type of sonar, called echolocation, to detect prey, avoid obstacles, and locate their roosting crevices in the dark. Bats emit sound pulses while flying and listen to their echoes from surrounding objects to assess their own location and those of the echoing objects (Yang and Gandomi, 2012).

Each pulse has a constant frequency (usually in the range of 25×10^3 to 150×10^3 Hz) and lasts a few thousandths of a second (up to about 8–10 ms). About 10–20 sounds are emitted every second with the rate of emission up to about 200 pps when they fly near their prey while hunting. If the interval between two successive sound bursts is less than 300–400 μs, bats cannot process them for path-finding purposes (Yang, 2010).

The speed of sound in air is typically $\nu = 340$ m/s, and the wavelength (W) of the ultrasonic sound bursts with a constant frequency (λ) is given by (Yang and Gandomi, 2012)

$$W = \frac{\nu}{\lambda} \tag{18.1}$$

in which W = the wavelength, ν = the speed of sound, and λ = frequency. W is in the range of 2–14 mm for the typical frequency range from 25×10^3 to 150×10^3 Hz. Such wavelengths W are of the same order of magnitude as their prey sizes.

Bats emit pulses as loud as 110 dB that are in the ultrasonic region (frequency range of human hearing is between 20 and 20 000 Hz). The loudness also varies from the loudest when searching for prey to a quieter base when homing toward the prey. The traveling range of such short pulses is typically a few meters.

Microbats can avoid obstacles as small as thin human hairs. Such echolocation behavior of microbats has been formulated to create the bat-inspired optimization algorithm applying the following idealized rules (Yang, 2010):

1) All bats use echolocation to sense distance, and they can discern the difference between food/prey and background barriers.
2) Bats fly randomly with velocity v_1 at position x_1 with a fixed frequency λ_{min}, varying wavelength W, and loudness A_0 to search prey. They can automatically adjust the wavelength (or frequency) of their emitted pulses and adjust the pulsation rate, depending on the proximity of their target.
3) The loudness can vary from a large (positive) A_0 to a minimum constant value A_{min}.

In general the frequency (λ) is in the range of $[\lambda_{min},\lambda_{max}]$ and corresponds to a range of wavelengths $[W_{min}, W_{max}]$. In actual implementations, one can adjust the range by adjusting the wavelengths (or frequencies), and the detectable range (or the largest wavelength) should be chosen such that it is comparable to the size of the domain of interest, and then toning down to smaller ranges. For simplicity, λ is assumed to be in the range of $[0,\lambda_{max}]$.

The pulsation rate (δ) is in the range of [0,1], where 0 means no pulses at all and 1 means the maximum pulsation rate. Based on these approximations and idealization, the basic steps of the BA have been summarized in the flowchart shown in Figure 18.1, and Table 18.1 lists the characteristics of the BA.

18.3 Creating an Initial Population

Each possible solution of the optimization problem represents a bat's position in the BA. A bat's position is defined by a set of N coordinates that constitute the decision variables. A bat's position is denoted by an array of size $1 \times N$ as follows:

$$Bat = X = \left(x_1, x_2, \ldots, x_i, \ldots, x_N\right) \tag{18.2}$$

where X = a solution (bat) of optimization problem, x_i = ith decision variable of solution X, and N = number of decision variables. The decision variable values $(x_1, x_2, x_3, \ldots, x_N)$ are represented as floating point numbers (real values) or as a predefined set for continuous and discrete problems, respectively.

The BA starts with the random generation (see Section 2.6) of a matrix of size $M \times N$ where M and N are the size of population and the number of decision variables, respectively. Hence, the matrix of solutions that is generated

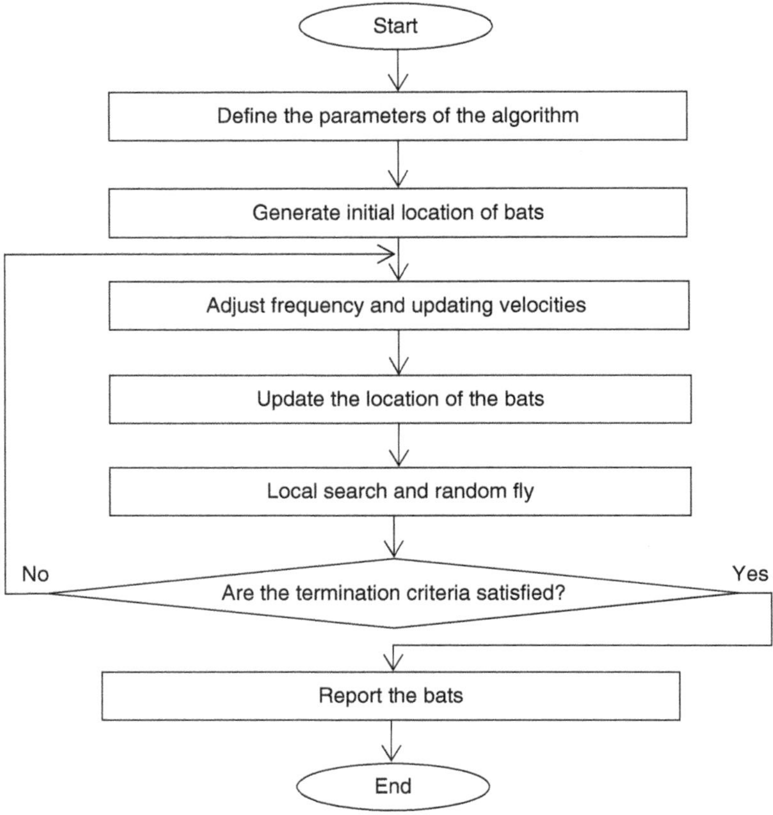

Figure 18.1 The flowchart of the BA.

Table 18.1 The characteristics of the BA.

General algorithm (see Section 2.13)	Bat algorithm
Decision variable	Position of bat in any dimension
Solution	Position of bat
Old solution	Old position of bat
New solution	New position of bat
Best solution	Best bat
Fitness function	Distance with food
Initial solution	Random bat
Selection	Loudness criteria
Process of generating new solutions	Fly bats

randomly is written as follows (rows and columns are the number of bats and the number of decision variables, respectively):

$$Population = \begin{bmatrix} X_1 \\ X_2 \\ \vdots \\ X_j \\ \vdots \\ X_M \end{bmatrix} = \begin{bmatrix} x_{1,1} & x_{1,2} & \cdots & x_{1,i} & \cdots & x_{1,N} \\ x_{2,1} & x_{2,2} & \cdots & x_{2,i} & \cdots & x_{2,N} \\ & & & \vdots & & \\ x_{j,1} & x_{j,2} & \cdots & x_{j,i} & \cdots & x_{j,N} \\ & & & \vdots & & \\ x_{M,1} & x_{M,2} & \cdots & x_{M,i} & \cdots & x_{M,N} \end{bmatrix} \tag{18.3}$$

in which X_j = jth solution, $x_{j,i}$ = ith decision variable of the jth solution, and M = population size.

18.4 Movement of Virtual Bats

According to Figure 18.1 the fitness value is evaluated for all solutions, and those solutions are ranked based on their fitness values. The following are the rules to update the jth bat's position (X_j) and its velocity in an N-dimensional search space ($j = 1, 2, ..., M$) (updated positions):

$$X_j^{(new)} = \left(x'_{j,1}, x'_{j,2}, \ldots, x'_{j,i}, \ldots, x'_{j,N}\right), \quad j = 1,2,\ldots,M \tag{18.4}$$

$$x'_{j,i} = x_{j,i} + v'_{j,i}, \quad j = 1,2,\ldots,M, \quad i = 1,2,\ldots,N \tag{18.5}$$

$$v'_{j,i} = v_{j,i} + \lambda_{j,i} \times \left(x_{j,i} - x_{Best,i}\right), \quad j = 1,2,\ldots,M, \quad i = 1,2,\ldots,N \tag{18.6}$$

$$\lambda_{j,i} = \left[\lambda_{min} + \left(\lambda_{max} - \lambda_{min}\right)\right] \times Rand, \quad j = 1,2,\ldots,M, \quad i = 1,2,\ldots,N \tag{18.7}$$

where $X_j^{(new)}$ = the new position of the jth bat (new solution), $x'_{j,i}$ = the new value of ith decision variable of the jth solution, $x_{j,i}$ = the old value of the ith decision variable of the jth solution, $Rand$ = a random value in the range of [0,1] drawn from a uniform distribution, $x_{Best,i}$ = ith decision variable of the current global best solution determined after comparing all the solutions among all the M bats, $v_{j,i}$ = the velocity of the jth bat (solution) in the ith dimension in the previous iteration, $v'_{j,i}$ = the velocity of the jth bat (solution) in the ith dimension in the current iteration, and λ_{min} and λ_{max} = lower and upper boundaries, respectively, of the frequency that are user-defined parameters of the algorithm and are determined based on the size of the decision space of the problem at hand. A frequency value is drawn from a uniform distribution in the range $[\lambda_{min}, \lambda_{max}]$ at the start of the BA and assigned to each bat.

18.5 Local Search and Random Flying

A new solution is locally generated using random walk once a solution has been selected among the current best solutions. This is the local search part (random fly) of the BA. The new solution that replaces the rth selected solution (Yang and Gandomi, 2012) is calculated as follows:

$$x'_{r,i} = x_{r,i} + Rnd(-1,1) \times A^{(t)}, \quad i = 1,2,\ldots,N \tag{18.8}$$

$$X_r^{(new)} = \left(x'_{r,1}, x'_{r,2}, \ldots, x'_{r,i}, \ldots, x'_{r,N}\right) \tag{18.9}$$

where $Rnd(-1,1)$ = a random number in the range of $[-1,1]$ and $A^{(t)}$ = the average loudness of all the bats at iteration t. The $A^{(t)}$ is reduced while approaching the optimum solution by using a rate called random walk rate.

Updating the velocities and positions of bats is similar to the procedure in the standard PSO algorithm as λ essentially controls the pace and range of the movement of the swarming particles. Therefore, the BA constitutes a combination of the standard PSO and intensive local search controlled by the loudness and pulsation rate.

18.6 Loudness and Pulse Emission

The BA implements local search when the pulse rate criterion is satisfied. The new solution replaces the old one when the loudness criterion is satisfied and the new solution is better than the old one. The pulse rate criterion is satisfied if a random value (*Rand*) is larger than δ_j. The loudness criterion is satisfied if a random value (*Rand*) is less than A_j. In addition, whenever the new solution replaces the old one, the loudness (A_j) and the pulsation rate (δ_j) are updated according to the BA's iteration steps. The loudness usually decreases once a bat has found its prey, while the pulsation rate increases. Thus, the loudness can be chosen as any value of convenience. For example, the values of $A^{(0)} = 1$ and $A_{min} = 0$ can be used, whereby the zero value means that a bat has just found a prey and temporarily stops emitting any sound. The pulsation and loudness rates at each iteration are calculated as follows:

$$\delta_j^{(t)} = \delta_j^{(Final)} \times \left(1 - e^{-C_1 \times t}\right), \quad j = 1,2,\ldots,M \tag{18.10}$$

$$A_j^{(t)} = C_2 \times A_j^{(t-1)} \tag{18.11}$$

where $\delta_j^{(t)}$ = pulsation rate of the jth solution at iteration t; $\delta_j^{(Final)}$ = final pulsation of the jth solution, which is a user-defined parameter; $A_j^{(t)}$ = loudness of the jth solution at iteration t; and C_1 and C_2 are constant values. C_2 is similar to the cooling factor in the simulated annealing (SA) algorithm. For any $0 < C_2 < 1$ and $C_1 > 0$, we have $A_j^{(t)} \to 0$ and $\delta_j^{(t)} \to \delta_j^{(Final)}$ when $t \to \infty$.

Choosing the correct values for the parameters C_1 and C_2 requires computational experimentation. Initially, each bat should be assigned values of loudness and pulsation rate. This can be achieved by randomization. Their loudness and pulsation rates are updated only if the solutions improve, which means that the bats are moving toward the optimal solution.

18.7 Termination Criteria

The termination criterion determines when to terminate the algorithm. Selecting a good termination criterion has an important role in the correct convergence of the algorithm. The number of iterations, the amount of improvement of solutions between consecutive iterations, and the run time are common termination criteria for the BA.

18.8 User-Defined Parameters of the BA

The population size (M), the initial loudness ($A^{(0)}$), the minimum loudness (A_{min}), the final pulsation rate ($\delta^{(Final)}$), the values of constant C_1 and C_2, the frequency boundaries (λ_{min} and λ_{max}), and the termination criteria are user-defined parameters of the BA. A good choice of the parameters depends on the decision space of a particular problem, and usually the optimal parameter setting for one problem is of limited utility for other problems. Determining a good set of parameter often requires performing computational experiments. A reasonable method for finding appropriate values for the parameters is performing sensitivity analysis, whereby combinations of parameters are tested and the algorithm is run several times for each combination to account for the random nature of the solution algorithm. In this manner, the analyst obtains a distribution of solutions and associated objective function values for each combination of parameters. A comparison of the results from all the combination of parameters provides guidance on a proper choice of the algorithmic parameters.

18.9 Pseudocode of the BA

```
Begin
  Input the parameters of the algorithm and initial data
  Generate M initial possible situations
  While (the termination criteria are not satisfied)
     Evaluate fitness value for all solutions
     Rank all solutions according to their fitness
       values and find the current best solution
```

```
    For j = 1 to M
        Generate new solutions by adjusting frequency,
          and updating velocities and locations/
          solutions
        Generate Rand randomly
        If Rand > δj
            Select a solution among the best
              solutions
            Generate a local solution around the best
              solution
        End if
        Generate a new solution by random fly
        Generate Rand randomly
        If (Rand < Aj) and (the new solution is
          better than the old one)
            Accept the new solutions
            Increase δj and reduce Aj
        End if
    Next j
  End while
  Report all solutions
End
```

18.10 Conclusion

This chapter described the BA, reviewed its development and applications, provided an analogy between the echolocation of bats and the BA, explained the BA in detail, and closed with a pseudocode for the BA.

References

Altringham, J. D. (1996). "Bats: Biology and behavior." Oxford University Press, Oxford.

Baziar, A., Kavoosi-Fard, A., and Zare, J. (2013). "A novel self adaptive modification approach based on bat algorithm for optimal management of renewable MG." Journal of Intelligent Learning Systems and Applications, 5(1), 11–18.

Bozorg-Haddad, O., Karimirad, I., Seifollahi-Aghmiuni, S., and Loáiciga, H. A. (2014). "Development and application of the bat algorithm for optimizing the operation of reservoir systems." Journal of Water Resources Planning and Management, 141(8), 04014097.

Colin, T. (2000). "The variety of life." Oxford University Press, Oxford.

Koffka, K. and Ashok, S. (2012). "A comparison of BA, GA, PSO, BP, and LM for training feed forward neural networks in e-learning context." International Journal of Intelligent Systems and Applications, 4(7), 23–29.

Malakooti, B., Kim, H., and Sheikh, S. (2012). "Bat intelligence search with application to multi-objective multiprocessor scheduling optimization." International journal of Advanced Manufacturing Technology, 60(9–12), 1071–1086.

Niknam, T., Sharifinia, S., and Azizipanah-Abarghooee, R. (2013). "A new enhanced bat-inspired algorithm for finding linear supply function equilibrium of GENCOs in the competitive electricity market." Journal of Energy Conversion and Management, 76, 1015–1028.

Ramesh, B., Mohan, V. C. J., and Ressy, V. C. V. (2013). "Application of bat algorithm for combined economic load and emission dispatch." Journal of Electrical Engineering and Telecommunications, 2(1), 1–9.

Reddy, V. U. and Manoj, A. (2012). "Optimal capacitor placement for loss reduction in distribution systems using Bat algorithm." IOSR Journal of Engineering (IOSRJEN), 10(2), 23–27.

Yang, X. S. (2010). "A new meta-heuristic bat-inspired algorithm." In: Cruz, C., Gonzalez, J. R., Pelta, D. A., and Terrazas, G. (Eds.), Nature inspired cooperative strategies for optimization (NISCO 2010), Studies in computational intelligence, Vol. 284, Springer, Berlin, Heidelberg, 65–74.

Yang, X. S. (2012). "Meta-heuristic optimization with applications: Demonstration via bat algorithm." Proceedings of 5th Bioinspired Optimization Methods and Their Applications (BIOMA2012), Bohinj, Slovenia, May 24–25, Ljubljana: Jozef Stefan Institute, 23–34.

Yang, X. S. and Gandomi, A. H. (2012). "Bat algorithm: A novel approach for global engineering optimization." Engineering Computations, 29(5), 464–483.

19

Plant Propagation Algorithm

Summary

This chapter describes the plant propagation algorithm (PPA) that emulates the multiplication of plants akin to strawberry. The history of the PPA and its applications are reviewed, the PPA is mapped to the natural process of plant propagation, the PPA is described in detail, and a pseudocode of the PPA is introduced.

19.1 Introduction

The plant propagation algorithm (PPA) is inspired by propagating plants akin to the strawberry plant (Salhi and Fraga, 2011). They tested the PPA with low-dimensional single- and multi-objective problems. The results showed that the PPA has merits and deserves further testing and research on higher-dimensional problems. Sulaiman et al. (2014) applied the PPA to solve large problems. The PPA is attractive because, among other things, it is easy to implement. It also involves only a few parameters that are relatively simple to specify unlike most other meta-heuristic algorithms.

19.2 Mapping the Natural Process to the Planet Propagation Algorithm (PPA)

The PPA resembles the manner in which plants, in particular strawberry plants, propagate. Although some varieties of plants propagate using seeds contained in fruits, hybrid types like strawberry are infertile and issue runners to propagate. In this way, the original plant issues runners to generate new plants.

Meta-Heuristic and Evolutionary Algorithms for Engineering Optimization, First Edition. Omid Bozorg-Haddad, Mohammad Solgi, and Hugo A. Loáiciga.

There is an interesting strategy that propagating plants employ. These plants develop runners. By doing so these plants attempt to place their offspring where nutrients and growth potential are suitable. If a plant is placed in a good spot of land, which provides enough water, nutrients, and light, it issues many short runners that generate new plants, which occupy the neighborhood as best as they can. However, if the original plant is placed in a spot without enough water, nutrients, light, or any other requirements called growth potential, it tries to find a better spot for its offspring. In the latter instance, the plant issues few runners farther to explore distant neighborhoods. It can be inferred that this plant sends only a few because sending a long runner is a large investment for a plant that is placed in a poor land.

The location of each plant represents a solution of the optimization problem in the PPA. The growth potential of the plant's location is synonymous to its fitness value. Generally, the richer the land, the better the fitness values. A plant propagation strategy is to generate new plants around itself using runners so that the number and length of runners are determined by the fitness value of the original (mother) plant. Runners represent the process by which new solutions of the optimization problem are generated. Table 19.1 lists the characteristics of the PPA.

The PPA consists of the following two critical rules:

1) Plants that are placed in appropriate lands propagate by spreading many short runners.
2) Plants that are placed in poor lands propagate by generating a few long runners.

Exploration and exploitation are important features of optimization algorithms. Exploration refers to the property of searching the space, while exploitation refers to the property of searching near good solutions for achieving a more precise solution. It is clear that in the PPA exploitation is executed by plants sending

Table 19.1 The characteristics of the PPA.

General algorithm (see Section 2.13)	Plant propagation algorithm
Decision variable	Position of plant in any dimension
Solution	Position of plant
Old solution	Mother plant
New solution	Daughter plant
Best solution	–
Fitness function	Growth potential
Initial solution	Random plant
Selection	Eliminating worst solutions
Process of generating new solutions	Propagation strategy

many short runners in good areas (high growth potential), while exploration is executed by sending few long runners by plants in poor areas.

The PPA starts by randomly generating a number of initial possible solutions (see Section 2.6) as plants within the decision space. The objective functions of all plants are evaluated, and the evaluated objective functions are normalized between zero and one. This allows ranking of all solutions according to their fitness values. In the next step, each plant acting as a mother plant generates daughter plants, which represent new solutions. This process is called propagation, and it is obvious that propagation proceeds according to the propagation strategy of plants. Plants with a strong fitness values generate more new solutions near themselves than those plants with inferior fitness values. Under maximization (minimization), strong (inferior) fitness value is tantamount to high (low) values of the fitness function. Each plant produces several offspring (new solutions), and, therefore, the population of solutions grows in each iteration. The worst solutions are eliminated at the end of each iteration to control the size of the population of solutions, and only a fixed number of solutions are kept and are carried to the next iteration. These solutions are considered as mother plants, and the aforementioned process is repeated until the termination criteria are satisfied. Figure 19.1 depicts the flowchart of the PPA.

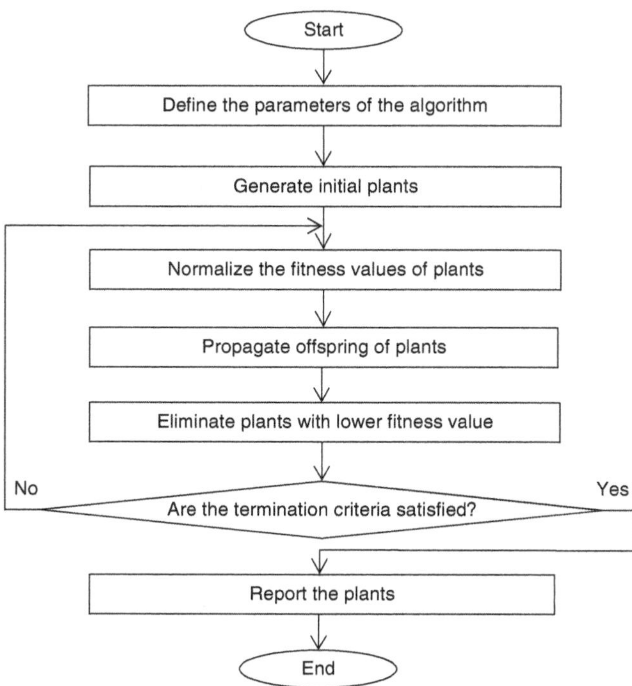

Figure 19.1 The flowchart of the PPA.

19.3 Creating an Initial Population of Plants

Each possible solution of the optimization problem represents a plant's position in the PPA. In an N-dimensional optimization problem, a plant's position is written as an array of size $1 \times N$, whose elements represent the decision variables. This array is defined as follows:

$$Plant = X = \left(x_1, x_2, \ldots, x_i, \ldots, x_N\right) \tag{19.1}$$

where X = a solution of the optimization problem, x_i = ith decision variable of solution X, and N = number of decision variables. The PPA algorithm starts by randomly generating a matrix of size $M \times N$ (see Section 2.6) where M and N are the size of the population and the number of decision variables, respectively. Hence, the matrix of solutions that is generated randomly is as follows (rows and columns represent the number of plants and the number of decision variables, respectively):

$$Population = \begin{bmatrix} X_1 \\ X_2 \\ \vdots \\ X_j \\ \vdots \\ X_M \end{bmatrix} = \begin{bmatrix} x_{1,1} & x_{1,2} & \cdots & x_{1,i} & \cdots & x_{1,N} \\ x_{2,1} & x_{2,2} & \cdots & x_{2,i} & \cdots & x_{2,N} \\ & & & \vdots & & \\ x_{j,1} & x_{j,2} & \cdots & x_{j,i} & \cdots & x_{j,N} \\ & & & \vdots & & \\ x_{M,1} & x_{M,2} & \cdots & x_{M,i} & \cdots & x_{M,N} \end{bmatrix} \tag{19.2}$$

in which X_j = jth solution, $x_{j,i}$ = ith decision variable of the jth solution, and M = population size. Each of the decision variable values $(x_1, x_2, x_3, \ldots, x_N)$ can be represented as floating point number (real values). The PPA solves problems with continuous decision spaces.

19.4 Normalizing the Fitness Function

The fitness of a solution must be assessed prior to generating new solutions. To accomplish this assessment, the fitness functions are normalized between zero and one. The following equation normalizes the fitness function:

$$\sigma\left[F\left(X_j\right)\right] = \frac{F\left(X_j\right) - Worst}{Best - Worst}, \quad j = 1, 2, \ldots, M \tag{19.3}$$

in which $\sigma[F(X_j)]$ = the normalized fitness value of the jth solution, $F(X_j)$ = fitness value of the jth solution, $Worst$ = the worst possible value of F, and $Best$ = the best possible value of F. Determining the best and worst possible values of the fitness value is sometimes impossible. In this case the best and worst available values in the current population or the best and worst fitness values calculated during the search process replace the best and worst possible values. Other functions such as trigonometric functions or exponential functions can be used for normalization depending on the problem at hand.

19.5 Propagation

The PPA dictates that each plant issues runners to explore the decision space. Each runner results in an offspring that is a new solution. The number of offspring made by a solution may vary among solutions. The number of runners (offspring) generated by a solution is proportionate to its normalized fitness value, and it is evaluated as follows:

$$\mu_j = \lceil \mu_{max} \times \sigma_j \times Rand \rceil, \quad j = 1,2,\ldots,M \tag{19.4}$$

in which μ_j = number of new solutions generated by the jth solution, μ_{max} = the maximum number of new solutions that can be produced by a solution (this is a predefined parameter), σ_j = the normalized fitness value of the jth solution, and $Rand$ = a random value from the range $[0,1]$. Notice that $\lceil x \rceil$ where x is the argument of the function on the right-hand side of Equation (19.4) means the ceiling of x (i.e., the smallest integer $\geq x$).

In contrast to the number of runners (new solutions), the length of runners is inversely proportional to the normalized fitness values. It was previously stated that better solutions generate new solutions close to themselves and poor solutions generate new solutions in places farther from themselves. The distance between the original solution and new solution is determined as follows:

$$d_{j,i} = 2 \times \left(1 - \sigma_j\right) \times (Rand - 0.5), \quad j = 1,2,\ldots,M, \quad i = 1,2,\ldots,N \tag{19.5}$$

in which $d_{j,i}$ = the length of runner of the jth solution in the ith dimension (decision variable). The term $\left(Rand - 0.5\right)$ makes it possible for a runner to take negative or positive directions.

The evaluated runner's length is employed to generate new solutions as follows:

$$x'_{r,i} = x_{j,i} + \left(x_i^{(U)} - x_i^{(L)}\right) \times d_{j,i}, \quad i = 1,2,\ldots,N, \quad r = 1,2,\ldots,\mu_j, \quad j = 1,2,\ldots,M \tag{19.6}$$

in which $x'_{r,i}$ = ith decision variable of the rth new solution generated by the jth solution, $x_{j,i}$ = ith decision variable of the jth solution, $x_i^{(U)}$ = the upper bound of the ith decision variable, and $x_i^{(L)}$ = the lower bound of the ith decision variable. Equation (19.6) may generate a new solution that falls outside the decision space.

In the latter instance, the newly generated solution is adjusted to be within the decision space. Each new solution r is represented as follows:

$$X_r^{(new)} = \left(x'_{r,1}, x'_{r,2}, \ldots, x'_{r,i}, \ldots, x'_{r,N}\right), \quad r = 1,2,\ldots,\mu_j \tag{19.7}$$

in which $X_r^{(new)}$ = rth new solutions generated by the jth solution.

19.6 Elimination of Extra Solutions

It is necessary at the end of each iteration to delete the part of population in excess of the allowed number of solutions. Solutions produce several offspring (new solutions) in each iteration, which means that the population would grow as the iterations progress unless the population is controlled in each iteration. This is achieved in each by eliminating the worst solutions to keep the number of solutions fixed as the algorithm progresses.

19.7 Termination Criteria

The termination criteria determine when to terminate the algorithm. Selecting a good termination criterion is important because if the number of iteration of the algorithm is not sufficiently large, the algorithm may terminate prematurely at a suboptimal solution. On the other hand, it is clear that wasteful computations are incurred if the algorithm continues to run when the solution does not improve across iterations. Although there are several distinct termination criteria, Salhi and Fraga (2011) recommended that the number of iterations is a suitable termination criterion. This means that the PPA algorithm runs for a predefined number of iterations.

19.8 User-Defined Parameters of the PPA

The population size (M), the maximum number of new solutions that can be produced by each solution (μ_{max}), and the maximum number of iterations are parameters that must be determined by the user. In comparison with other meta-heuristic and evolutionary algorithms, it is seen that the PPA has a relatively small number of user-defined parameters. Its simple structure and small number of parameters make implementation of the PPA comparatively simple. A good choice of the parameters depends on the decision space of a particular problem, and usually the optimal parameter setting for one problem is of limited utility for any other problem. Determining a good set of parameters often requires performing computational experiments. A reasonable method for finding appropriate values for the parameters is performing sensitivity analysis, whereby combinations of parameters are tested and the algorithm is run several times for each combination to account for the random nature of the solution algorithm. In this manner, the analyst obtains a distribution of solutions and associated objective function values for each combination of parameters. A comparison of the results from all the combination of parameters provides guidance on a proper choice of the algorithmic parameters.

19.9 Pseudocode of the PPA

```
Begin
  Input the parameters of the algorithm and initial data
  Let M = the size of population and N = number of
    decision variables
  Generate M initial possible solutions randomly
  While (the termination criteria are not satisfied)
     Evaluate fitness value of solutions
     For j = 1 to M
         Evaluate normalized fitness value of solution
           j (σ[F(X_j)])
         Evaluate number of new solutions generated by
           solution j (μ_j)
         For i = 1 to N
            Evaluate the length of runner d_j,i
         Next i
         For r = 1 to μ_j
            For i = 1 to N
                Evaluate the decision variable i-th of
                  new solution r-th (x'_r,i)
                If x'_r,i > x_i^(U) or x'_r,i < x_i^(L)
                     Adjust x'_r,i within the boundaries
                End if
            Next i
         Next r
     Next j
     Constitute new population with M best solutions
  End while
  Report the population
End
```

19.10 Conclusion

This chapter described the PPA that simulates multiplication of some plants akin to the strawberry plant. The chapter presented a brief introduction to the PPA, its analogy to plant propagation, an algorithmic explanation, and a pseudocode of the PPA.

References

Salhi, A. and Fraga, E. S. (2011). "Nature-inspired optimization approaches and the new plant propagation algorithm." Proceedings of the International Conference on Numerical Analysis and Optimization (ICeMATH 2011), Yogyakarta, Indonesia, June 6–8, Colchester: University of Essex.

Sulaiman, M., Salhi, A., Selamoglu, B. I., and Kirikchi, O. B. (2014). "A plant propagation algorithm for constrained engineering." Mathematical Problems in Engineering, 2014, 627416.

20

Water Cycle Algorithm

Summary

This chapter describes the water cycle algorithm (WCA), which is a relatively new meta-heuristic optimization algorithm. The fundamental concepts of the WCA are inspired by natural phenomena concerning the water cycle and how rivers and streams flow to the sea. The next sections present the background and applications of the WCA, explain the WCA, and provide a pseudocode.

20.1 Introduction

The water cycle algorithm (WCA) was introduced by Eskandar et al. (2012). The authors compared the results of the WCA with those of other meta-heuristic algorithms such as the genetic algorithm (GA), particle swarm optimization (PSO) algorithm, harmony search (HS), bee colony, and differential evolution (DE). Their results showed that the WCA is a suitable method for solving constrained optimization problems and competes favorably with other meta-heuristic algorithms. Eskandar et al. (2013) illustrated the application of the WCA by solving the problem of designing truss structures and compared the results with those of other meta-heuristic algorithms such as the GA, PSO, mine blast algorithm (MBA), etc. The results of their comparison demonstrated the strong capability of the WCA algorithm to find optimal solutions and its rapid convergence. Bozorg-Haddad et al. (2014) applied the WCA to find optimal operation strategies for a four-reservoir system in Iran. The results demonstrated the high efficiency and reliability of the WCA in solving reservoir operation problems. Ghaffarzadeh (2015) applied the WCA to design a power system stabilizer (PSS) that enhances the damping of power system oscillations.

Meta-Heuristic and Evolutionary Algorithms for Engineering Optimization,
First Edition. Omid Bozorg-Haddad, Mohammad Solgi, and Hugo A. Loáiciga.

20.2 Mapping the Water Cycle Algorithm (WCA) to the Water Cycle

The basic idea of the WCA is inspired by nature's water cycle and by the manner in which rivers and streams flow toward the sea. The water or hydrologic cycle has no beginning or end, and its several processes occur consecutively and indefinitely. Streams and rivers constitute interconnected networks, issuing from high ground, where sometimes snow or ancient glaciers melt, and discharging to sea and lakes. Streams and rivers collect water from rain and other streams on their way down stream toward the sea. Water in rivers, seas, and lakes evaporates. The evaporated water is carried to the atmosphere to generate clouds. These clouds condense and release the water back in the form of rain or snow, creating streams and rivers. This is the manner of functioning of the hydrologic or water cycle (see, e.g., David, 1993).

The WCA simulates the precipitation process by randomly generated raindrops, each of which is an array that represents a solution of the optimization problem. The streams are created by the raindrops, and streams join each other to form rivers. Some of the streams may also flow directly to the sea. Rivers and streams flow to the sea (the lowest point). The WCA classifies raindrops as the sea, or as rivers, or as streams that form an interconnected network. The sea is the best raindrop (solution), which has the minimum fitness value (under minimization), and other raindrops are known as rivers or streams, so that rivers are better solutions than streams. Rivers flow to the sea, and streams flow to rivers or to the sea. The WCA generates new solutions as water flows toward the sea. The evaporation process causes the seawater to evaporate as rivers/streams flow to the sea. Whenever all rivers have fitness values as good as that of the sea, this indicates that all the water has evaporated and raining occurs again completing the water cycle. Table 20.1 lists the characteristics of the WCA.

The WCA assumes that there is rain or precipitation that generates raindrops (initial solutions) randomly. The fitness values of all the raindrops are evaluated following precipitation. The best raindrop, which has the lowest value of objective function (under minimization), is marked out as the sea, and other raindrops are classified into rivers and streams according to their fitness values. In the next step, the number of streams connected to each river is determined according to the fitness of each river. In fact, each river receives water from the streams depending on its flow magnitude. New streams and rivers (new solutions) are generated by old streams flowing to their corresponding rivers and by rivers flowing to the sea. The direction of flow is reversed if new streams are better than the old corresponding rivers. In other words, a new stream that is better than an old river becomes a river, and an old river becomes a stream. Also, the direction of flow between rivers and the sea can be reversed if a new river is better than an old sea. The flow of water through streams and rivers toward the sea continues until all rivers reach the

Table 20.1 The characteristics of the WCA.

General algorithm (see Section 2.13)	Water cycle algorithm
Decision variable	Position of raindrop in any dimension
Solution	Raindrop/sea/river/stream
Old solution	Sea/river/stream
New solution	New place of stream
Best solution	Sea
Fitness function	Flow
Initial solution	Random stream
Selection	Categorizing streams into sea/river/stream
Process of generating new solutions	Stream and river flow

sea, which indicates that evaporation has been completed, at which time rain (or precipitation) occurs again to form a new network of streams. This algorithm is repeated until the termination criteria are satisfied. Figure 20.1 presents the flowchart of the WCA.

20.3 Creating an Initial Population

The WCA designates possible solution of the optimization problem as a raindrop. In an N-dimensional optimization problem, a raindrop is an array of size $1 \times N$. This array is defined as follows:

$$Raindrop = X = (x_1, x_2, \ldots, x_i, \ldots, x_N) \tag{20.1}$$

where X = a solution of optimization problem, x_i = ith decision variable of solution X, and N = number of decision variables. To start the optimization algorithm, a matrix of size $M \times N$ is generated (where M and N are the size of population and the number of decision variables, respectively). Hence, the matrix of solutions that is generated randomly (see Section 2.6) is written as follows (rows and columns are the number of raindrops and the number of decision variables, respectively):

$$Population = \begin{bmatrix} X_1 \\ X_2 \\ \vdots \\ X_j \\ \vdots \\ X_M \end{bmatrix} = \begin{bmatrix} x_{1,1} & x_{1,2} & \cdots & x_{1,i} & \cdots & x_{1,N} \\ x_{2,1} & x_{2,2} & \cdots & x_{2,i} & \cdots & x_{2,N} \\ & & & \vdots & & \\ x_{j,1} & x_{j,2} & \cdots & x_{j,i} & \cdots & x_{j,N} \\ & & & \vdots & & \\ x_{M,1} & x_{M,2} & \cdots & x_{M,i} & \cdots & x_{M,N} \end{bmatrix} \tag{20.2}$$

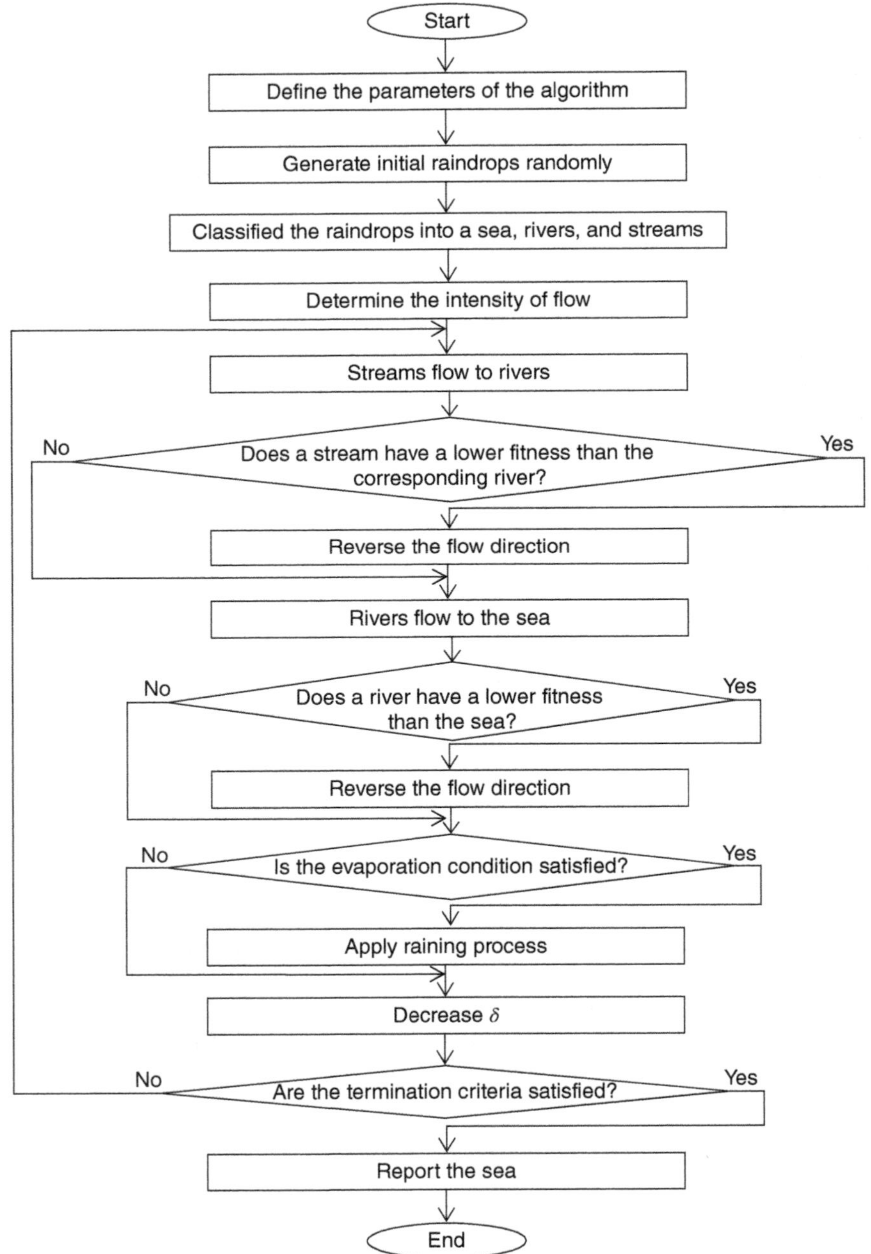

Figure 20.1 The flowchart of the WCA for a minimization problem.

in which $X_j = j$th solution, $x_{j,i} = i$th decision variable of the jth solution, and M = population size. Each of the decision variable values $(x_1, x_2, x_3, \ldots, x_N)$ can be represented as floating point number (real values) or as a predefined set for continuous and discrete problems, respectively.

20.4 Classification of Raindrops

The raindrop that has the minimum fitness value among others is marked out as the sea after evaluating the objective function of all the solutions. A number R of the best raindrops are selected as rivers. The total number of streams S that flow to the rivers or may directly flow to the sea is calculated as follows:

$$S = M - R - \underbrace{1}_{sea} \tag{20.3}$$

in which R = the total number of rivers and S = the total number of streams.

Figure 20.2 illustrates how to classify raindrops and the relations between streams, rivers, and raindrops.

The following equations are used to designate/assign raindrops to the rivers and sea depending on the intensity of the flow (fitness value) in a minimization problem:

$$\lambda_{sea} = Round\left\{\left|\frac{F(Sea)}{F(Sea) + \sum_{r=1}^{R} F(River_r)}\right| \times S\right\} \tag{20.4}$$

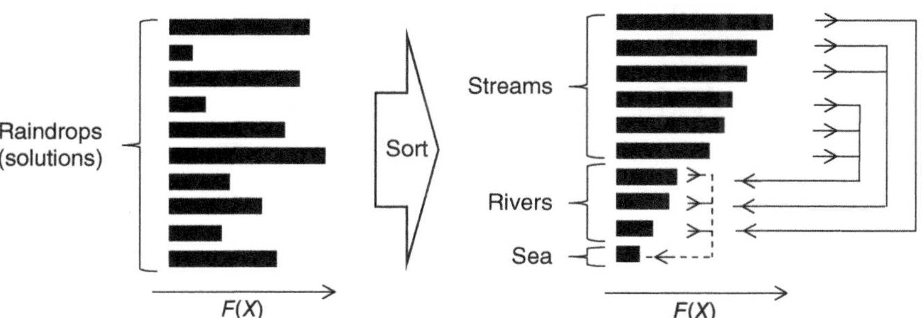

Figure 20.2 Classification of raindrops and relations between the sea, rivers, and streams according to their fitness values ($F(X)$) for a minimization problem where $M = 10$ and $R = 3$.

$$\lambda_j = Round\left\{\left|\frac{F(River_j)}{F(Sea)+\sum_{r=1}^{R}F(River_r)}\right|\times S\right\},\quad j=1,2,\ldots,R \tag{20.5}$$

where $F(X)$ = fitness value of a solution X, $Round$ = a function that rounds off the value of the function within brackets to the closest integer number, Sea = the best solution, $River_j$ = jth good solution after the best solution, λ_{sea} = the number of streams that flow directly to the sea, and λ_j = the number of streams that flow to the jth river.

20.5 Streams Flowing to the Rivers or Sea

The streams are created from the raindrops and join each other to form new rivers. Some of the streams may also flow directly to the sea. All rivers and streams discharge to the sea (best optimal point). A stream flows to a river along a line connecting them using a randomly chosen distance calculated as follows:

$$X' = Rnd(0, C\times d),\quad C>1 \tag{20.6}$$

where X' = new stream, Rnd = a random value between 0 and $(C\times d)$, d = the current distance between old stream and river, and C = a value between 1 and 2 (near to 2). The best value for C may be chosen as 2. The value of C being greater than one enables streams to flow in different directions toward the rivers. Figure 20.3 shows the schematic of a stream's flow toward a specific river.

The concept behind Equation (20.6) involving the flow of streams to rivers may also be used for the flow of rivers to the sea. Therefore, the new decision variables for new streams and rivers are as follows:

$$Stream_{j,i}^{(new)} = Stream_{j,i} + Rand\times C\times\left(River_i^{(j)} - Stream_{j,i}\right)\qquad i=1,2,\ldots,N,\quad j=1,2,\ldots,S \tag{20.7}$$

$$River_{j,i}^{(new)} = River_{j,i} + Rand\times C\times\left(Sea_i - River_{j,i}\right)\qquad i=1,2,\ldots,N,\quad j=1,2,\ldots,R \tag{20.8}$$

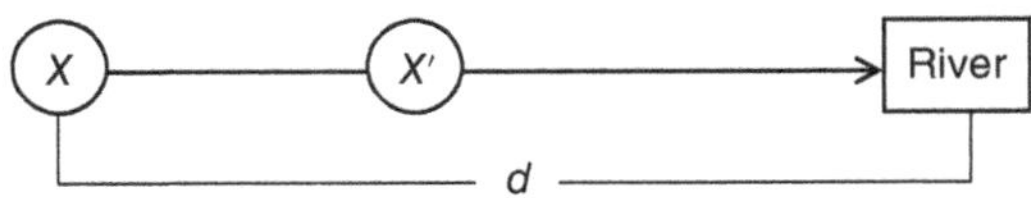

Figure 20.3 Schematic of stream flowing toward a river at distance d: X: existing stream; X': new stream.

where $Stream_{j,i}^{(new)}$ = new value of the ith decision variable of jth stream, $Stream_{j,i}$ = the old value of the ith decision variable of the jth stream, $River_i^{(j)}$ = the value of the ith decision variable of the river connected to the jth stream, $River_{j,i}^{(new)}$ = new value of the ith decision variable of the jth river, $River_{j,i}$ = the old value of the ith decision variable of the jth river, Sea_i = the value of the ith decision variable of the sea (best solution), and $Rand$ = uniformly distributed random number in the range [0,1].

The positions of a river and stream are exchanged (i.e., the stream becomes a river and vice versa) whenever the solution given by a stream is better than that of the river to which it discharges after calculation of the new stream and river. Such exchange can similarly happen for rivers and the sea.

20.6 Evaporation

Evaporation is one of the most important factors that can prevent the WCA algorithm from rapid convergence (immature convergence). As can be observed in nature, water evaporates from rivers and lakes. The evaporated water is carried to the atmosphere to form clouds that then condenses in the colder atmosphere, releasing the water back to earth in the form of rain (or snow, in which case we deal with snow pellets as equivalents to rain drops). The rain creates the new streams and rivers, and the water cycle continues. The WCA induces evaporation from seawater as rivers/streams flow to the sea. This assumption is necessary to avoid entrapment in local optima. The following commands show how to determine whether or not a river flows to the sea:

$$d = \left|Sea - River_j\right|, \quad j = 1,2,\ldots,R \tag{20.9}$$

where d = the distance between the sea and rivers. If the distance between a river and the sea is less than a predefined threshold δ ($d < \delta$), this indicates that the river has reached/joined the sea. In this instance, the evaporation process is applied, and, as seen in nature, the precipitation will start after sufficient evaporation has occurred. Therefore, a large value of δ reduces the search, while a small value encourages the search intensity near the sea. δ controls the search intensity near the sea (the optimal solution). It is recommended that δ be chosen as a small number close to zero (Eskandar et al., 2012). The value of δ gradually decreases as the algorithm progresses as follows:

$$\delta^{(t+1)} = \delta^{(t)} - \frac{\delta^{(t)}}{T} \tag{20.10}$$

where T = the total number of algorithm's iterations and $\delta^{(t)}$ = the threshold of evaporation in iteration t.

20.7 Raining Process

The raining process is applied after satisfying the evaporation process. New raindrops are randomly generated during the raining process. The decision variables of the new raindrops are generated as follows:

$$x_i^{(new)} = Rnd\left(x_i^{(L)}, x_i^{(U)}\right), \quad i = 1,2,\ldots,N \tag{20.11}$$

where $x_i^{(new)}$ = decision variable ith of a new raindrop (solution) and $x_i^{(U)}$ and $x_i^{(L)}$ = lower and upper bounds defined by the given problem, respectively. Again, the best newly formed raindrop is considered as a river flowing to the sea. The other new raindrops are assumed to form new streams that flow to the rivers or may flow directly to the sea.

The convergence rate and computational performance of the WCA for constrained problems is improved by applying Equation (20.12) to generate the decision variable of new streams that flow directly to the sea. This equation encourages the generation of streams that flow directly to the sea to improve the search near the sea (the optimal solution) in the feasible region of constrained problems:

$$Stream_i^{(new)} = Sea_i + \sqrt{\eta} \times Randn, \quad i = 1,2,\ldots,N \tag{20.12}$$

where $Stream_i^{(new)}$ = ith decision variable of the new stream, η = coefficient that shows the range of the search region near the sea, and $Randn$ = normally distributed random number. A larger value for η increases the possibility to exit from the feasible region. On the other hand, the smaller value for η leads the algorithm to search in smaller regions near the sea. A suitable value for η is 0.1. From a mathematical viewpoint, the term $\sqrt{\eta}$ represents the standard deviation, and, accordingly, η captures the concept of variance. Using these concepts, the generated individuals with variance η are distributed about the best obtained optimal point (the sea).

There are several differences between the WCA and other meta-heuristic methods such as the PSO. The WCA treats rivers (a number of best selected points except the best one (the sea)) as "guidance points," which guide other individuals in the population toward better positions to prevent searching in inappropriate regions near optimal solutions. Furthermore, rivers are not fixed points and they flow toward the sea (the best solution). This procedure (streams flowing to rivers and rivers flowing to the sea) leads to search moves toward the best solution. In contrast, the PSO prescribes that only individuals (particles) find the best solution based on their best personal experiences. The WCA also uses "evaporation and raining conditions" that resemble the mutation operator in the GA. The evaporation and raining conditions prevent the WCA algorithm from being trapped in local solutions (Eskandar et al., 2012).

20.8 Termination Criteria

The termination criterion determines when to terminate the algorithm. Selecting a good termination criterion has an important role in the correct convergence of the algorithm. The number of iterations, the amount of improvement of the solution between consecutive iterations, and the run time are common termination criteria for the WCA.

20.9 User-Defined Parameters of the WCA

The population size (M), the number of rivers (R), the initial threshold of evaporation (δ), and the termination criteria are user-defined parameters of the WCA. A good choice of the parameters depends on the decision space of a particular problem, and usually the optimal parameter setting for one problem is of limited utility for any other problem. Determining a good set of parameter often requires performing computational experiments. A reasonable method for finding appropriate values for the parameters is performing sensitivity analysis, whereby combinations of parameters are tested and the algorithm is run several times for each combination to account for the random nature of the solution algorithm. In this manner, the analyst obtains a distribution of solutions and associated objective function values for each combination of parameters. A comparison of the results from all the combination of parameters provides guidance on a proper choice of the algorithmic parameters.

20.10 Pseudocode of the WCA

```
Begin
  Input the parameters of the algorithm and initial data
  Generate M initial possible solutions randomly
  Evaluate fitness value for all solutions
  Classified solutions into streams, rivers and the
    sea and assign each stream to a river or the sea
  While (the termination criteria are not satisfied)
     For j = 1 to S (total number of streams)
         Flow stream j toward the corresponding river
           or the sea
         If the new generated stream is better than
           the river or the sea
             Reverse the flow direction
         End if
```

```
        Next j
        For j = 1 to R (total number of rivers)
            Flow river j toward the sea
            If the new generated river is better than the sea
                Reverse the flow direction
            End if
        Next j
        If evaporation condition is satisfied
            Start raining process
        End if
        Reduce the value of δ
    End while
    Report the best solution (the sea)
End
```

20.11 Conclusion

This chapter described the WCA, a meta-heuristic optimization algorithm. The chapter presented a brief history of the development and applications of the WCA and described the analogy between the water cycle and the mathematical statement of the WCA. The chapter also described the WCA in detail and introduced a pseudocode.

References

Bozorg-Haddad, O., Moravej, M., and Loáiciga, H. A. (2014). "Application of the water cycle algorithm to the optimal operation of reservoir systems." Journal of Irrigation and Drainage Engineering, 141(5), 04014064.

David, S. (1993). "The water cycle, illustrations by John Yates." Thomson Learning, New York.

Eskandar, H., Sadollah, A., and Bahreininejad, A. (2013). "Weight optimization of truss structures using water cycle algorithm." International Journal of Civil Engineering, 3(1), 115–129.

Eskandar, H., Sadollah, A., Bahreininejad, A., and Hamdi, M. (2012). "Water cycle algorithm-a novel metaheuristic optimization method for solving constrained engineering optimization problems." Computer and Structures, 110–111, 151–166.

Ghaffarzadeh, N. (2015). "Water cycle algorithm based power system stabilizer robust design for power systems." Journal of Electrical Engineering, 66(2), 91–96.

21

Symbiotic Organisms Search

Summary

This chapter describes the symbiotic organisms search (SOS) algorithm, a recently developed meta-heuristic algorithm. Unlike most of meta-heuristic algorithms, the SOS does not require specification of algorithmic parameters. First, the basic concepts of the SOS algorithm are mapped to the symbiotic relations among organisms. The steps of the SOS algorithm are defined in detail and a pseudocode of the SOS is presented.

21.1 Introduction

Cheng and Prayogo (2014) introduced the symbiotic organisms search (SOS) algorithm. The SOS is a nature-inspired optimization algorithm that simulates three different symbiosis interactions within a paired organism relationship through an ecosystem. Evolutionary algorithms (EAs) are targets of criticism because of the need for specifying algorithmic parameters. The SOS algorithm requires only the specification of the "maximum number of evaluations" and the "population size." Evi et al. (2015) implemented the SOS for solving capacitated vehicle routing problem (CVRP). Rajathy et al. (2015) demonstrated the superiority of the SOS for solving economic load dispatch problem.

21.2 Mapping Symbiotic Relations to the Symbiotic Organisms Search (SOS)

Symbiosis is a close physical relation between two interacting organisms. There are three categories of symbiotic relationships including mutualism, commensalism, and parasitism. Mutualism is a relation that is beneficial to

Meta-Heuristic and Evolutionary Algorithms for Engineering Optimization, First Edition. Omid Bozorg-Haddad, Mohammad Solgi, and Hugo A. Loáiciga.

both organisms involved. In many mutualistic relationships, the relation is obligatory; the species cannot live without each other. In others, the species can exist separately but are more successful when they are involved in a mutualistic relation. For instance, the interaction between starlings and buffalo is known as mutualism. Starlings remove ticks from the buffalo's skin for sustenance. The itching from ticks biting the buffalo is reduced in return. Commensalism takes place when an organism receives benefits while the other is not significantly affected. Such is the case of the interactions between remora fish and sharks. Remora fish eat leftovers from the shark without bothering the shark at all. Parasitism is another kind of symbiotic relation in which an organism obtains benefits from the interaction while other is harmed. One example is that of the anopheles mosquito biting humans for blood. Anopheles injects *Plasmodium* parasites into the human body that cause malaria, a potentially lethal disease.

The SOS algorithm does not reproduce or create children (or offspring), a trait that differentiates it from GA-type EAs (Cheng and Prayogo, 2014). It does, however, like the majority of population-based EAs, generate an initial population of solutions (called "ecosystem") plus specific operators through an iterative process to search for a near-optimal solution among a group of candidate solutions (called "organisms") within the promising area of the search space.

The SOS algorithm simulates the ecosystem with a randomly generated set of solutions, each of which is known as an organism. The solutions or organisms are represented as an array of decision variables of the optimization problem. Commonly, meta-heuristics have operators that generate a new solution. The phases in the SOS such as mutualism, commensalism, and parasitism serve as the operators. Each organism interacts with other organisms randomly in the population through all phases. When simulating mutualism, both of two selected solutions are improved. The simulation of commensalism of two selected solutions improves one solution while leaving the other one unchanged. Parasitism is simulated when an improved solution replaces another solution that is discarded (it dies). The features of the SOS algorithm are listed in Table 21.1. Figure 21.1 illustrates the steps of the SOS algorithm.

21.3 Creating an Initial Ecosystem

Each possible solution of the optimization problem is called an organism in the SOS. An organism or solution is represented as array of size $1 \times N$ in an N-dimensional optimization problem. This array is written as follows:

$$Organism = X = \left(x_1, x_2, \ldots, x_i, \ldots, x_N\right) \tag{21.1}$$

Table 21.1 The characteristics of the SOS.

General algorithm (see Section 2.13)	Symbiotic organisms search
Decision variable	Elements of organism
Solution	Organism
Old solution	Old organism
New solution	Improved organism
Best solution	Best organism
Fitness function	Quality of organism
Initial solution	Random organism
Selection	Selection of organism
Process of generating new solutions	Symbiotic relationship

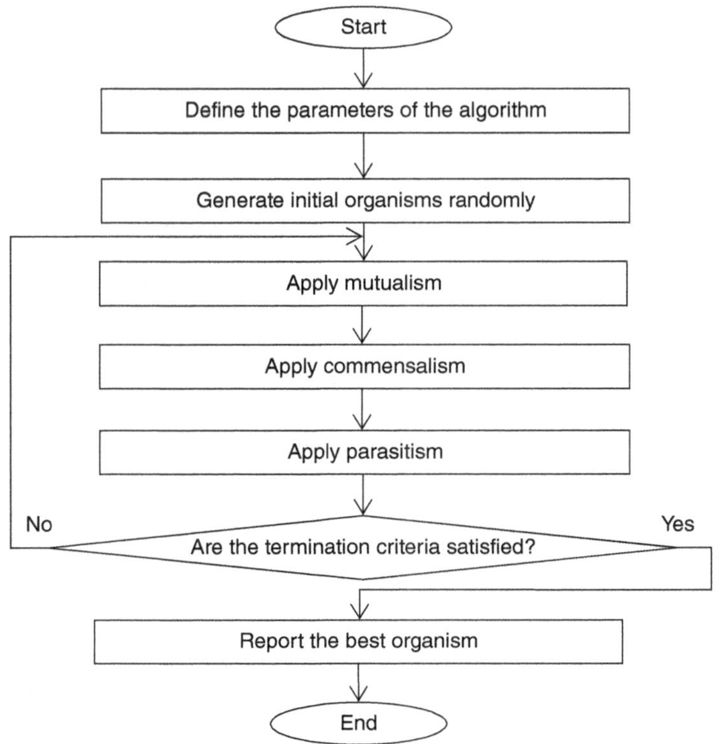

Figure 21.1 The flowchart of the SOS.

where X = a solution of optimization problem, x_i = ith decision variable of solution X, and N = number of decision variables. The SOS algorithm starts with the generation of matrix of size $M \times N$ (where M and N are the size of population and the number of decision variables, respectively). Hence, the matrix of solutions that is generated randomly (see Section 2.6) is written as follows (rows and columns are the number of organisms and the number of decision variables, respectively):

$$Population = \begin{bmatrix} X_1 \\ X_2 \\ \vdots \\ X_j \\ \vdots \\ X_M \end{bmatrix} = \begin{bmatrix} x_{1,1} & x_{1,2} & \cdots & x_{1,i} & \cdots & x_{1,N} \\ x_{2,1} & x_{2,2} & \cdots & x_{2,i} & \cdots & x_{2,N} \\ & & & \vdots & & \\ x_{j,1} & x_{j,2} & \cdots & x_{j,i} & \cdots & x_{j,N} \\ & & & \vdots & & \\ x_{M,1} & x_{M,2} & \cdots & x_{M,i} & \cdots & x_{M,N} \end{bmatrix} \tag{21.2}$$

in which X_j = jth solution, $x_{j,i}$ = ith decision variable of the jth solution, and M = population size. Each of the decision variable values $(x_1, x_2, x_3, \ldots, x_N)$ can be represented as floating point number (real values) or as a predefined set for continuous and discrete problems, respectively.

21.4 Mutualism

In this phase two organisms participate in a dialectic relation that is beneficial for both of them. Let X_j be an organism representing the jth member of the ecosystem (i.e., the population of solutions), and the rth solution X_r is selected randomly from the ecosystem to interact with X_j. Sometimes X_j and X_r are located far from each other in the search space. Thus, providing a mechanism to explore some new regions within the search space would promote the search algorithm's performance. In so doing mutualism evaluates the new candidate solutions with a mutual factor to update (improve) the two organisms simultaneously rather than separately. The mutual factor is given by the following formula:

$$\mu_i = \frac{x_{j,i} + x_{r,i}}{2}, \quad i = 1,2,\ldots,N \tag{21.3}$$

in which μ_i = mutual factor in the ith dimension, $x_{j,i}$ = ith decision variable of the jth solution of the population of solutions (or ecosystem), and $x_{r,i}$ = ith decision variable of the rth solution of the population of solutions. The new candidate solutions are obtained with the following formulas:

$$x'_{j,i} = x_{j,i} + Rand \times \left(x_{Best,i} - \mu_i \times \beta_1\right), \quad i = 1,2,\ldots,N \tag{21.4}$$

$$x'_{r,i} = x_{r,i} + Rand \times \left(x_{Best,i} - \mu_i \times \beta_2 \right), \quad i = 1,2,\ldots,N \tag{21.5}$$

$$X_j^{new} = \left(x'_{j,1}, x'_{j,2}, \ldots, x'_{j,i}, \ldots, x'_{j,N} \right) \tag{21.6}$$

$$X_r^{new} = \left(x'_{r,1}, x'_{r,2}, \ldots, x'_{r,i}, \ldots, x'_{r,N} \right) \tag{21.7}$$

where $x'_{j,i}$ = new value of the ith decision variable of the jth solution, $x'_{r,i}$ = new value of the ith decision variable of the rth solution, $Rand$ = a random value in the range [0,1], $x_{Best,i}$ = ith decision variable of the best solution in the population, X_j^{new} = a candidate solution for the jth solution, X_r^{new} = a candidate solution for the rth solution, and β_1 and β_2 = constants. β_1 and β_2 are valued randomly as either 1 or 2 to reflect the level of benefits received from the symbiosis. The constants β_1 and β_2 are so chosen to reflect the fact that in a mutualistic symbiosis between two organisms, one organism might gain a large benefit, while the other might receive no significant benefit.

21.5 Commensalism

By definition commensalism is a relation between two organisms whereby one organism gains benefits while the other remains unaffected. Similar to the previous section, X_r is selected randomly from the ecosystem to interact with X_j; however, X_j strives to receive benefits from the relation, yet X_r remains neutral or unaffected. In this phase, a new candidate solution that may outperform X_j is calculated as follows:

$$x'_{j,i} = x_{j,i} + Rand \times \left(x_{Best,i} - x_{r,i} \right), \quad i = 1,2,\ldots,N \tag{21.8}$$

$$X_j^{new} = \left(x'_{j,1}, x'_{j,2}, \ldots, x'_{j,i}, \ldots, x'_{j,N} \right) \tag{21.9}$$

21.6 Parasitism

The SOS algorithm has a unique mutation operator called parasitism in which X_j and X_r are the artificial parasite and host, respectively. In this type of symbiosis relation, one organism benefits, while the other is harmed. The trademark of the parasite vector (PV) is that it competes against other randomly selected organisms rather than with its parent/creator. Throughout this phase the PV attempts to replace X_r, which is selected randomly from the ecosystem. So as to create a PV, X_j must be duplicated within the search space, and then, the random dimensions are modified by using random numbers. Specifically, let $X_j = (x_1, x_2, \ldots, x_i, \ldots, x_N)$ be a randomly selected solution, and assume the ith decision variable is randomly selected for modification.

Parasitism calculates a new solution $X_j^{(parasitism)} = (x_1, x_2, \ldots, x_i', \ldots, x_N)$ in which x_i' is evaluated as follows:

$$x_i' = Rnd\left(x_i^{(L)}, x_i^{(U)}\right) \tag{21.10}$$

in which x_i' = value of ith decision variable of $X_j^{(parasitism)}$ and $x_i^{(L)}$ and $x_i^{(U)}$ = the lower and upper boundary of the ith decision variable, respectively.

The fitness values of the organisms illustrate the superiority of the parasite or the host. If $X_j^{(parasitism)}$ outperforms X_r, then it will remain in the ecosystem and X_r is deleted. Otherwise, X_r remains in the ecosystem (i.e., the population of solutions).

21.7 Termination Criteria

The termination criterion determines when to terminate the algorithm. Selecting a good termination criterion has an important role in the correct convergence of the algorithm. The number of iterations, the amount of improvement of the solution between consecutive iterations, and the run time are common termination criteria for the SOS.

21.8 Pseudocode of the SOS

```
Begin
  Input the parameters of the algorithm and initial data
  Generate M initial possible solutions randomly
  Evaluate fitness value for all solutions
  While (the termination criteria are not satisfied)
     Determine the best solution (Best) in population
       according to the fitness values
     For j = 1 to M
         Select organism r (X_r) from ecosystem randomly
         Generate X_j^(new) and X_r^(new) by mutualism and
           evaluate their fitness values
         If (X_j^(new) is better than X_j) and (X_r^(new) is better
           than X_r)
             Replace the new solutions for previous
               solutions
         End if
         Select organism r (X_r) from ecosystem randomly
         Generate X_j^(new) based on commensalism and
           evaluate its fitness value
         If X_j^(new) is better than X_j
```

```
            Replace the new solution for previous
              solution
        End if
        Duplicate Xj (Xj(new) = Xj)
        Modify Xj(new) based on parasitism and evaluate
          its fitness value
        Select organism r (Xr) from ecosystem
          randomly
        If Xj(new) is better than Xr
             Replace Xj(new) for Xr (Xr = Xj(new)).
        End if
    Next j
  End while
  Report the best solution
End
```

21.9 Conclusion

The SOS algorithm is a recently developed meta-heuristic algorithm that, unlike most of meta-heuristic algorithms, does not require specification of algorithmic parameters. This chapter described the SOS algorithm, presented its analogy to symbiosis, reported the SOS's algorithmic steps, and closed with a pseudocode.

References

Cheng, M. Y. and Prayogo, D. (2014). "Symbiotic organisms search: A new metaheuristic optimization algorithm." Computers and Structures, 139(1), 98–112.

Evi, R., Yu, V. F., Budi, S., and Redi, A. A. N. P. (2015). "Symbiotic organism search (SOS) for solving the capacitated vehicle routing problem." International Journal of Mechanical, Aerospace, Industrial, Mechatronic and Manufacturing Engineering, 9(5), 857–861.

Rajathy, R., Tarawinee, B., and Suganya, S. (2015). "A novel method of using symbiotic organism search algorithm in solving security-constrained economic dispatch." International Conference on Circuit, Power and Computing Technologies (ICCPCT), Nagercoil, India, March 19–20, Piscataway, NJ: Institute of Electrical and Electronics Engineers (IEEE).

22

Comprehensive Evolutionary Algorithm

Summary

This chapter introduces a new meta-heuristic optimization algorithm called the comprehensive evolutionary algorithm (CEA). This algorithm combines and takes advantages of some aspects of various algorithms, especially the genetic algorithm (GA) and the honey-bee mating optimization (HBMO) algorithm. The following sections describe the fundamentals of the CEA and its algorithmic details. The chapter closes with a pseudocode of the CEA.

22.1 Introduction

The comprehensive evolutionary algorithm (CEA) is an optimization algorithm of recent vintage that combines features of the genetic algorithm (GA) and the honey-bee mating optimization (HBMO) algorithm. The CEA can solve single and multi-objective problems. This algorithm optimizes the defined objective function of an optimization problem based on three processes: (1) selection, (2) production, and (3) replacement. In addition, the CEA is able to explicitly perform sensitivity analysis of some of its parameters based on the problem conditions. In general, the CEA has better convergence performance and speed to the near-optimal solution, on the optimality of final solution, and on the run time period.

The GA was developed by Holland (1975), inspired by evolutionary process that are emulated mathematically in the GA. Numerous researches have been carried out to improve, extend, and apply the GA to a wide variety of optimization problems (Dimou and Koumousis, 2003; Hormwichian et al., 2009; Sonmez and Uysal, 2014). The HBMO algorithm is a population-based method for optimization in which the searching process for finding the optimal solution is inspired by honey-bee mating. Bozorg-Haddad et al. (2006)

Meta-Heuristic and Evolutionary Algorithms for Engineering Optimization,
First Edition. Omid Bozorg-Haddad, Mohammad Solgi, and Hugo A. Loáiciga.

developed the HBMO algorithm, evaluated its performance, and compared it with other algorithms for solving several mathematical problems and a simple reservoir operation problem. The HBMO algorithm has been applied to various optimization problems with continuous and discrete decision spaces and has shown acceptable results in solving them (Bozorg-Haddad et al., 2008, 2010, 2011; Solgi et al., 2015; Bozorg-Haddad et al., 2016; Solgi et al., 2016a). Solgi et al. (2016b) modified the HBMO, leading to the enhanced HBMO (EHBMO) algorithm, and demonstrated the superiority of the EHBMO compared with the HBMO and the elitist GA in solving several problems.

22.2 Fundamentals of the Comprehensive Evolutionary Algorithm (CEA)

CEA is based on the main concepts of the GA and the HBMO algorithm and implements a wide range of selection and generation operators that are selectively applied by the user in the optimization process to solve optimization problems based on the user's choice.

The CEA employs various operators. Selection operators select the superior solutions among the existing ones in each evolution step or iteration. Generation operators produce new solutions based on existing ones. The selection process refers to selecting some solutions to generate new solutions. The fitness function of selected solutions must be superior among the current iteration solutions. This implies that the probability of improvement will increase in the next iteration, and it can be expected that the algorithm would advance correctly toward a solution using various selection operators. The CEA features four selection operators: (1) proportionate, (2) tournament, (3) random, and (4) Boltzmann selection operator. The first three operators are germane to the GA, while the fourth to the HBMO algorithm. All four operators can be used in the CEA. Generating new solutions in the CEA is performed by crossover and mutation processes. The exclusive feature of the CEA is that it identifies efficient operators during the problem solution and relies on them for continuing the optimization procedure. Also, the CEA takes advantage of elitism, which defines a process in which the best solutions of the previous iteration are carried to the next iteration without any change. The CEA also ensures that the best solution produces a significant part of the next population by applying the Boltzmann selection operator. In this respect the CEA resembles some features of the HBMO. However, the CEA is not restricted to a limited number of operators and it employs several selection and generation operators, even those of the GA. The CEA can therefore apply various selection and generation operators to rapidly reach a near-optimal solution. These characteristics are improvements of the CEA not present in previous algorithms.

Many of parameters in the CEA are defined as decision variables and as dependent parameters on the characteristics of the problem. This feature

Table 22.1 The characteristics of the CEA.

General algorithm (see Section 2.13)	Comprehensive evolutionary algorithm
Decision variable	Gene of chromosome
Solution	Chromosome (individual)
Old solution	Parent
New solution	Children (offspring)
Best solution	Elite
Fitness function	Quality of individual
Initial solution	Random individual
Selection	Selection
Process of generating new solution	Reproduction

bypasses the need of conducting sensitivity analysis of its parameters. Parameters' values are determined based on problem conditions during the optimization process. The operators that exhibit better performance during the optimization process are recognized in each iteration and their effect is logically increased in the next iterations proportional to their performance. Thus, the effect of operators with poor performance in the optimization process is gradually decreased, but these operators are not completely removed from the optimization process. Lastly, the user can assess the impact of the applied operators in each optimization problem and identify those that are efficient in solving specific types of problems. In summary, sensitivity analysis of its parameters is implicitly performed by the CEA itself.

The CEA treats each solution as an individual (chromosome). Each chromosome is constructed of genes that represent decision variables. The fitness values of individuals determine their quality. Offspring or children, which represent new solutions, are generated by genetic operators including crossover and mutation. Also, the best individuals in the population of solutions in each iteration are known as elites. Table 22.1 lists the characteristics of the CEA.

The CEA starts receiving the input information including algorithmic parameters (estimates) and other necessary input data associated with the problem at hand. The values of some of these parameters are determined by the user, and other values are calculated by the algorithm based on the optimization problem's characteristics during the optimization procedure. After receiving the input information, the CEA initializes the portion of operators according to the problem characteristics and randomly generates initial possible solutions in the allowable range of the problem. The CEA calculates the fitness value of the initial solutions, and it starts its iterative calculations by applying a trial-and-error procedure to find the near-optimal solution. After selecting the best solutions (elites) among the existing ones, the selection operators select superior solutions according to the operators' roles. New solutions are generated by applying the selected superior solutions. This is accomplished by the crossover and mutation operators. These new solutions are employed in the next iteration. The fitness values of new solutions are calculated. The CEA calculates the improvement that its algorithmic operators effect on the possible solutions and compares the values of the fitness functions of the new solutions with that of the best one. At this time, the operators' roles in the next iteration are modified proportionally to the amount of improvement in the fitness function achieved by them in the current iteration, thus completing the iteration of the algorithm. The algorithm stops executing and reports the final solution when the stopping criteria are satisfied. The flowchart of the CEA is shown in Figure 22.1.

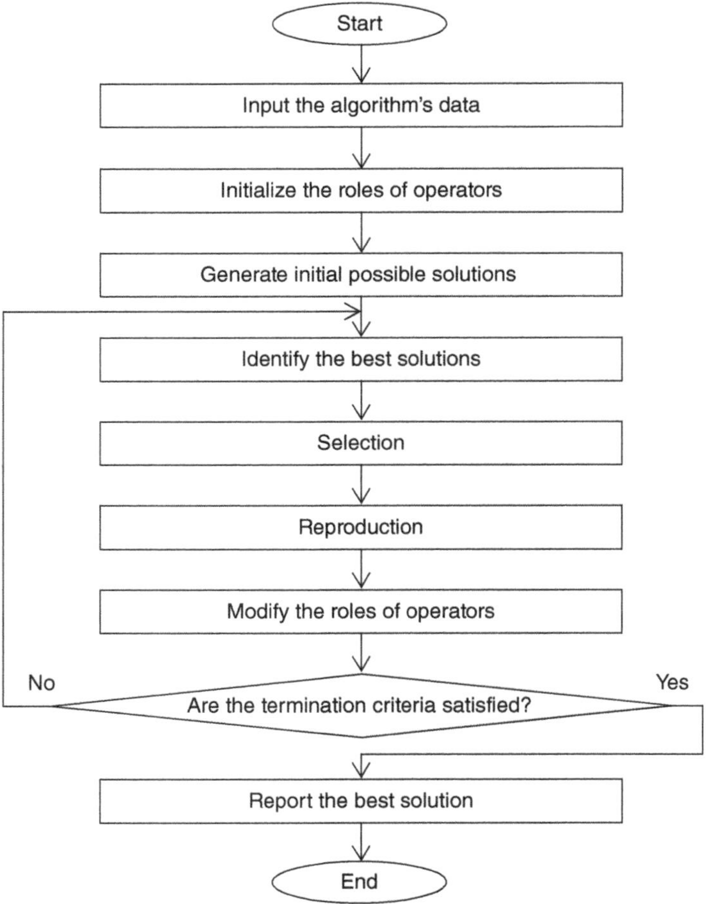

Figure 22.1 The flowchart of the CEA.

All the evolutionary algorithms have similarities to that shown in Figure 22.1. Yet, the CEA is more comprehensive than other algorithms from the viewpoint of selection and generation operators. The CEA is able to identify efficient operators to objectively apply them in the optimization. Also, the CEA carries out sensitivity analysis of its parameters automatically.

22.3 Generating an Initial Population of Solutions

The CEA calls each possible solution of the optimization problem an individual. Each individual symbolizes a series of gens (decision variables) that constitutes a solution of the problem in the mathematical formulation of an optimization problem. In an N-dimensional optimization problem, an individual is represented by an array of size $1 \times N$. This array is defined as follows:

$$Individual = X = \left(x_1, x_2, \ldots, x_i, \ldots, x_N\right) \tag{22.1}$$

where X = a solution of optimization problem, x_i = ith decision variable of solution X, and N = number of decision variables. In the CEA, the decision variable values $(x_1, x_2, x_3, \ldots, x_N)$ are real values.

The CEA starts by randomly generating a matrix of size $M \times N$ where M and N are the size of population and the number of decision variables, respectively. Hence, the matrix of solutions that is generated randomly (see Section 2.6) is written as follows (rows and columns are the number of individuals and the number of decision variables, respectively):

$$Population = \begin{bmatrix} X_1 \\ X_2 \\ \vdots \\ X_j \\ \vdots \\ X_M \end{bmatrix} = \begin{bmatrix} x_{1,1} & x_{1,2} & \cdots & x_{1,i} & \cdots & x_{1,N} \\ x_{2,1} & x_{2,2} & \cdots & x_{2,i} & \cdots & x_{2,N} \\ & & & \vdots & & \\ x_{j,1} & x_{j,2} & \cdots & x_{j,i} & \cdots & x_{j,N} \\ & & & \vdots & & \\ x_{M,1} & x_{M,2} & \cdots & x_{M,i} & \cdots & x_{M,N} \end{bmatrix} \tag{22.2}$$

in which X_j = jth solution, $x_{j,i}$ = ith decision variable of the jth solution, and M = population size.

22.4 Selection

Selection is the procedure by which some individuals are chosen from the current population or decision space for reproduction. There are different selection operators. Applying four different selection operators ((1) proportionate selection, (2) tournament, (3) random, and (4) Boltzmann selection operator) is

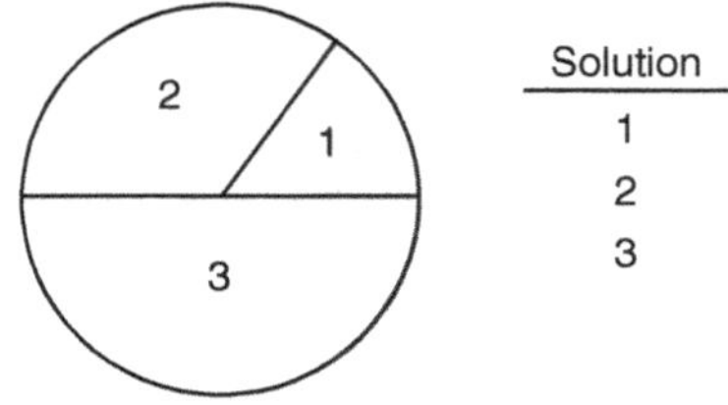

Solution	F	P
1	60	0.50
2	40	0.33
3	20	0.17

Population size $(M) = 3$

Figure 22.2 Illustration of a roulette wheel.

possible in the CEA, and the user can activate anyone of or all of them for solving an optimization problem. As mentioned previously, the three first operators are applied in the GA, while the fourth one is used in the HBMO algorithm. All four can be used with the CEA.

Applying the proportionate selection operator requires normalization of the solution fitness functions in each iteration based on their summation. These normalized values are considered as the selection probability of each solution. The selection probability of solutions with better fitness value exceeds those of undesirable solutions when applying the proportionate selection operator. The selection probability of less desirable solutions is not zero. Proportionate selection computes the probability of an individual being selected as follows (under maximization):

$$P_k = \frac{F(X_k)}{\sum_{j=1}^{M} F(X_j)} \tag{22.3}$$

in which P_k = the probability of the kth solution being selected and $F(X_k)$ = the fitness function of solution X_k.

The probability of selection of each solution is evaluated. A solution k has a chance proportionate to P_k to be selected. Based on the evaluated probabilities, a roulette wheel is made and turned to select solutions. The concept of a roulette wheel is depicted in Figure 22.2, using a trivial example with a population of three solutions. Each individual (solution) possesses a part of a roulette wheel that is proportionate to its fitness value. The selection is random and any individual has a chance to be selected. Clearly selection favors the fitter individuals on average (see Section 4.4).

The tournament selection operator selects randomly two or more solutions from the current population of solutions. The best solution among the selected ones is added to the list of selected solutions. This procedure is repeated as many times as needed (see Section 4.4). The selected solution in each step can remain in the population and may be selected again in the next steps or it can removed from the population in which case it will not be selected again.

The latter procedure is considered in the CEA to select different solutions with this operator and to avoid selecting similar solutions in each iteration. Solutions that are not selected remain in the current population.

Another selection operator that is employed in the CEA is random selection. This operator selects solutions randomly among the current population. Applying this operator can cause generating a random and uniform set of solutions in the next iteration. This affects the optimization procedure of algorithm and the convergence speed of the CEA algorithm might be decreased.

The CEA also applies the Boltzmann selection operator, which selects superior solutions by generating new random ones that are compared with the best one in the current iteration and are selected if they are superior to the current best. If the newly random generated solution is not better than the best one in the current iteration, the Boltzmann probability function is calculated as follows:

$$P = \exp\left(\frac{-\left|Best - F(X)\right|}{\mu_0 \times (1-\alpha)^{\beta}}\right) \tag{22.4}$$

$$\beta = 1 + \frac{100 \times t}{T} \tag{22.5}$$

in which $Best$ = fitness value of the best solution in the current iteration (elite); X = a randomly generated solution; $F(X)$ = fitness function of random generated solution (X); μ_0 = initial index of time elapsed since the start of the algorithm implementation, whose value fall within a specific range; α = random coefficient in the range of (0,1); t = number of the current iteration; and T = total number of iterations of the CEA algorithm.

P in Equation (22.4) is the selection probability of a newly generated random solution. It is compared with a random value in the range of [0,1]. If the random value is less than P, the newly generated solution is selected. Otherwise, another a new solution will be generated and evaluated (see Section 12.5).

22.5 Reproduction

A thorough search of the decision space in each optimization problem is possible with new solutions obtained through the reproduction process. The CEA classifies generating operators as crossover and mutation operators.

22.5.1 Crossover Operators

Crossover is a process in which a new solution is generated using two solutions selected from the current iteration. Table 22.2 lists the crossover operators that are employed in the CEA. In Table 22.2 a section is divided into left, middle, and right sides by vertical lines or section cuts, as shown in Figures 22.3 and 22.4.

Table 22.2 Types of crossover operators in the CEA.

Crossover operator	Type	Definition
One-point cut	(1)	The left side of the cross section of the first selected solution is crossed over with the right side of the cross section of the second selected solution
	(2)	The right side of the cross section of the first selected solution is crossed over with the left side of the cross section of the second selected solution
	(3)	The left side of the cross section of the first selected solution is crossed over uniformly with the right side of the cross section
	(4)	The left side of the cross section of the second selected solution is crossed over uniformly with the right side of the cross section
	(5)	The right side of the cross section of the first selected solution is crossed over uniformly with the left side of the cross section
	(6)	The right side of the cross section of the second selected solution is crossed over uniformly with the left side of the cross section
	(7)	The left side of the cross section of the first selected solution is crossed over with fixed weighted crossover for all of decision variables on the right side of the cross section
	(8)	The right side of the cross section of the first selected solution is crossed over with fixed weighted crossover with all the decision variables on the left side of the cross section
	(9)	The left side of the cross section of the first selected solution is crossed over with variable weighted crossover with all the decision variables on the right side of the cross section
	(10)	The right side of the cross section of the first selected solution is crossed over with variable weighted crossover with all the decision variables on the left side of the cross section
Two-point cut	(11)	The middle part of cross sections of the first selected solution is crossed over with the sides of the cross sections of the second selected solution
	(12)	The middle part of cross sections of the second selected solution is crossed over with the sides of the cross sections of the first selected solution
	(13)	The sides of the cross sections of the first selected solution are crossed over with uniform crossover in the middle part of the cross sections
	(14)	The sides of the cross sections of the second selected solution are crossed over with uniform crossover in the middle part of cross sections

Table 22.2 (Continued)

Crossover operator	Type	Definition
	(15)	The middle part of the cross sections of the first selected solution is crossed over uniformly with the sides of the cross sections
	(16)	The middle part of the cross sections of the second selected solution is crossed over uniformly with the sides of the cross sections
	(17)	The sides of the cross sections of the first selected solution are crossed over with fixed weighted crossover with all the decision variables on the middle part of the cross sections
	(18)	The middle part of the cross sections of the first selected solution is crossed over with fixed weighted crossover with all the decision variables on the sides of the cross sections
	(19)	The sides of the cross sections of the first selected solution are crossed over with variable weighted crossover with all of decision variables on the middle part of cross sections
	(20)	The middle part of the cross sections of the first selected solution is crossed over with variable fixed weighted crossover with all the decision variables on the sides of cross sections
Whole crossover	(21)	Uniform whole crossover
	(22)	Fixed weighted whole crossover for all decision variables in both solutions' structure
	(23)	Variable weighted whole crossover for all decision variables in both solutions' structure

Various types of crossover operators in the CEA are classified into three general categories: (1) One-point cut crossover in which only one section cut is considered in the structure of selected solutions to generate new ones (Figure 22.3), (2) two-point cut crossover in which two section cuts are considered in the structure of selected solutions to generate new ones (Figure 22.4), and (3) overall crossover in which the whole set of selected solutions is considered to generate new solutions without any section cut (Figure 22.5). It was previously stated that a row of decision variables is considered as a possible solution of the problem (see Equation (22.2)). Therefore, in one- and two-point cut crossovers, one and two sections, respectively, are assumed in the structure of a solution. There is not section cut in the whole crossover operators.

In the first new solution of Figure 22.3b (Figure 22.3c), the left [right] side of the section cut (depicted by a vertical line) is related to the first selected solution. The crossover on the right [left] side of the section cut is done uniformly.

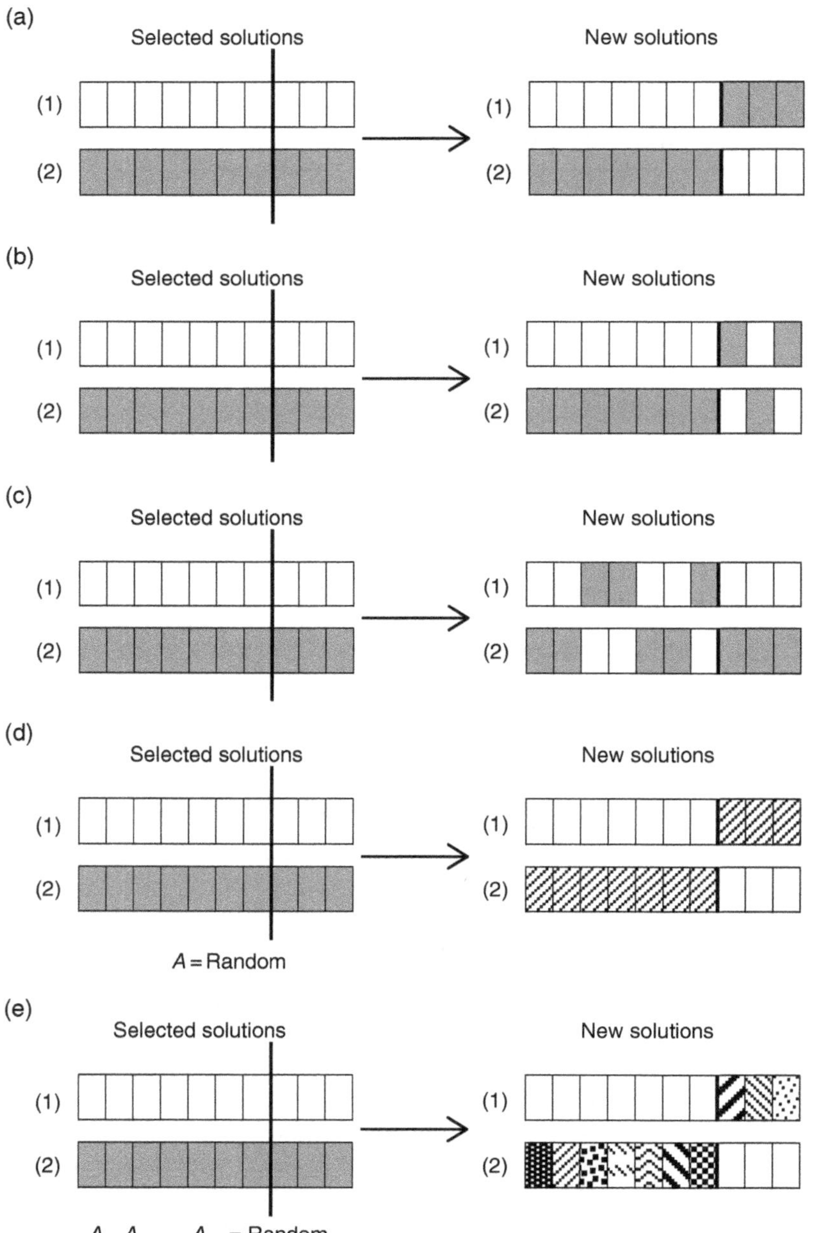

Figure 22.3 Types of one-point cut crossover in the CEA: (a) Type 1 (new solution (1)) and type 2 (new solution (2)), (b) Type 3 (new solution (1)) and type 4 (new solution (2)), (c) Type 5 (new solution (1)) and type 6 (new solution (2)), (d) Type 7 (new solution (1)) and type 8 (new solution (2)), and (e) Type 9 (new solution (1)) and type 10 (new solution (2)).

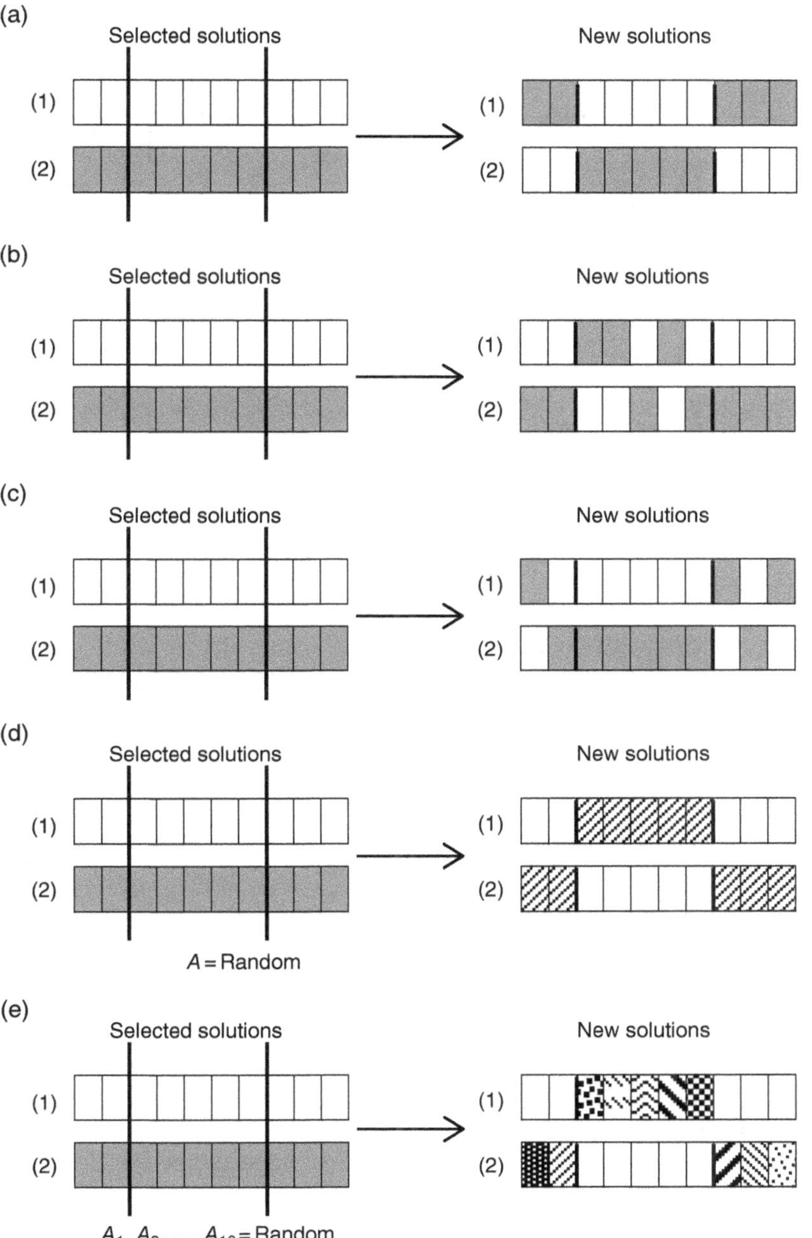

Figure 22.4 Types of two-point cut crossover in the CEA: (a) Type 11 (new solution (1)) and type 12 (new solution (2)), (b) Type 13 (new solution (1)) and type 14 (new solution (2)), (c) Type 15 (new solution (1)) and type 16 (new solution (2)), (d) Type 17 (new solution (1)) and type 18 (new solution (2)), and (e) Type 19 (new solution (1)) and type 20 (new solution (2)).

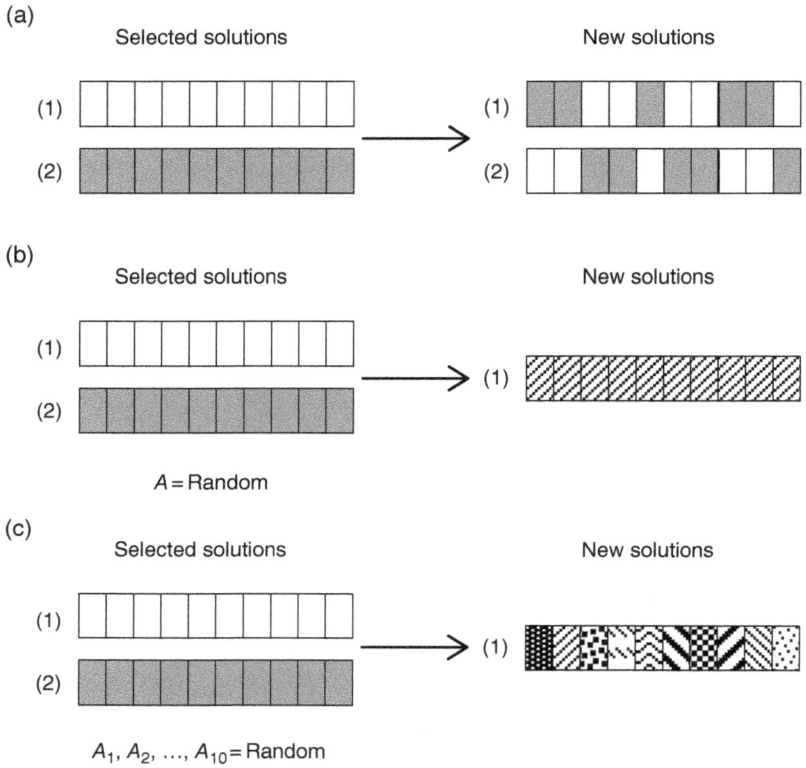

Figure 22.5 Types of overall crossover in CEA: (a) Type 21, (b) Type 22, and (c) Type 23.

This means that each decision variable on the right [left] side of the section cut of the new solution may be related to the first or second solution, and it determined randomly with the same chance for both solutions. In the first and second new solutions of Figure 22.3d (Figure 22.3e), the left and right sides of the section cut are related to the first selected solution, respectively. The crossover in another side of the section cut is performed by weighting. This means that a random value for all decision variables (a new random value for each decision variable) (A) is considered in the range of [0,1]. Then, the right side of the section cut is determined as the sum of the product of A times the first selected solution plus the product of $1 - A$ times the second one (the first new solution).

In the first new solution of Figure 22.4b (Figure 22.4c), the two sides of the section cuts are (middle of section cuts is) related to the first selected solution. The crossover in the middle of section cuts (two sides of section cuts) is done uniformly. This means that each decision variable in the middle of section cuts (two sides of section cuts) of the new solution may be related to the first or second solution and it determined randomly with the same chance for both solutions. In the first and second new solutions of Figure 22.4d (Figure 22.4e), the two sides and the middle of section cuts are related to the first selected solution, respectively. The crossover in the middle and two sides of section cuts is performed by weighting considering a random value for all the decision variables (a new random value for each decision variable) (A) in the range of [0,1]. The middle of the section cuts is determined as sum of the product of A times the first selected solution plus the product of $1-A$ times the second one (the first new solution).

The new solutions of Figure 22.5a undergo overall crossover that is done uniformly. This means that each decision variable of the new solution may be related to the first or second solution, and it determined randomly with the same chance for both solutions. In Figure 22.5b (Figure 22.5c), the overall crossover is performed by weighting considering a random value for all decision variables (a new random value for each decision variable) (A) in the range of [0,1]. The structure of a new solution is determined as sum of the product of A times the first selected solution plus the product of $1-A$ times the second solution.

It should be noted that the performance of each new solution of the various crossover operators is assessed separately by the CEA. Also, the 23 types of crossover operators employed by the CEA perform differently and search thoroughly the decision space of an optimization problem. It may be possible that the allowable range constraints are not satisfied in the types of crossovers shown in Figures 22.3d, e, 22.4d, e, and 22.5b, c. In this case a random solution in the allowable range for decision variables is generated to replace any generated infeasible solution.

22.5.2 Mutation Operators

Mutation is a process that generates new solutions in the next iteration or improves solutions generated by the crossover operator in evolutionary algorithms. It also expands the searching for solutions by the CEA algorithm in the decision space. Different types of mutation operators applied by in the CEA are listed in Table 22.3.

Four types of mutation operators are considered in the CEA: (1) Random mutation involves parts of the selected solution structure that are randomly mutated in the allowable range and a new solution is obtained. (2) Boundary mutation change the parts of the selected solution structure, which may be closer to the upper, middle, and lower bounds of the allowable range in the problem, through boundary mutation, which is done based on the upper and lower bounds of the allowable range of the decision variables, obtaining a new solution. (3) Directional mutation change parts of selected solution structure

Table 22.3 Types of mutation operators in the CEA.

Type	Definition
(1)	Random mutation (randomly changes some parts of the selected solution's structure in the allowable range)
(2)	Boundary mutation (makes the structure of the selected solution closer to the boundaries of the allowable range)
(3)	Directional mutation based on gradient of the fitness function of the selected solution is compared with the fitness function of the best solution
(4)	Dynamic mutation based on the upper and lower boundaries of the allowable range (the values of 0 and 1 mean mutation based on the upper and lower boundaries, respectively)

through oriented mutation, also obtaining a new solution. In this case, the gradient of fitness function for a selected solution is calculated and compared with the fitness function of the best solution in the set. Mutation is performed so that the calculated new solution becomes closer to the best one. (4) Dynamic mutation dynamically mutates parts of the selected solution structure based on the upper and lower bounds of the allowable range.

These mutation operators applied by the CEA have different performances, and their solutions differ from each other. A random solution is generated in the allowable range of the decision variables that replaces an infeasible solution whenever the allowable range constraints are not satisfied in the third and fourth types of the mutation operator.

22.6 Roles of Operators

Various selection and generation operators are selected by the user to solve an optimization problem. An operator role is to measure the percentage of the total number of solutions that is produced by that operator. In other words, it is the number of solutions selected or generated by each operator for the next iteration. The operator roles are changed during the optimization according to their performance, and it renders the number of existing solutions in each iteration variable. A key factor in generating operators is their probability of occurrence, which measures the frequency of application of each operator.

The roles of operators are updated by evaluating the performance of various selection and production operators after calculating the fitness function of newly obtained solutions. In other words, the fitness values of new solutions are calculated, and the improvement caused by each operator is calculated and they are compared with the best solution. The operators' roles in the next iteration are modified proportionally to the amount of improvement of the fitness function they created in the current iteration. The operators that have better performance during the optimization process are identified in each iteration, and their effect is increased in the next iteration proportional to their performance. Also, the effect of operators that have not desirable performance in the optimization process is gradually decreased, but these operators are not completely removed from the optimization process. Therefore, the CEA assesses directly the effect of different operators in the optimization process and identifies the efficient operators for each problem. This implies that an implicit sensitivity analysis is performed for applying selected operators for different processes in the CEA by the user. The effect of operators that have no desirable performance is reduced to improve the quality and convergence speed of the optimization.

22.7 Input Data to the CEA

Input data to the CEA includes algorithmic parameters and data for the problem's simulation model. Some of these parameters are determined by the user, while the others are determined by the CEA. Table 22.4 lists the inputs and their determination procedure in the CEA. As shown in this table, the first five algorithmic parameters and all of the simulation model parameters are determined by the user of the CEA. The simulation model parameters are determined according to problem conditions and they usually do not require sensitivity analysis.

The first algorithmic parameter is the number of runs. Evolutionary algorithms generate a set of random solutions. It is therefore necessary to evaluate several algorithmic runs to assess the quality of the calculated solutions. The CEA can perform several runs in parallel and present the final solution of each run individually. Applying the CEA requires determining the number of runs as specified by the user. The optimization process in the CEA starts with a set of initial solutions. Thus, the number of solutions in the initial population is specified by the user, while the suitable range for selecting the value of this parameter is determined by the algorithm based on the problem conditions, and it is indicated to the user. The necessity of sensitivity analysis for this parameter is decreased because the algorithm assists the user in selecting a suitable population size. The CEA applies all of the operators automatically for solving a problem in case the user selects none of the selection and generation operators. The CEA can identify the best operators and objectively apply them during the optimization process.

Table 22.4 List of the parameters of the CEA.

Parameter		Determined by
Algorithm parameters	(1) Number of runs	User
	(2) Number of algorithm iterations	
	(3) Desired precision of calculations	
	(4) Number of solutions in the primary population	
	(5) Type of selection and generation operators	
	(6) Number of solutions considered as elite	Algorithm (CEA)
	(7) Portion of selection and generation operators	
	(8) Probability of generation operators	
Information of the problem simulation model	(1) Number of objective functions	User
	(2) Optimization status of each objective function (maximization or minimization)	
	(3) Number of decision variables	
	(4) Allowable range for decision variable values	
	(5) Coefficients of calculating and controlling the problem constraints	

Parameters (6)–(8) in Table 22.4 are automatically determined by the algorithm according to the problem characteristics, and their sensitivity analysis is implicitly done in the algorithm. The number of elites, which are the best solutions of the previous iteration that are carried to next solution without any change, is determined by the algorithm. Thus, the necessity of sensitivity analysis for this parameter vanishes in the CEA.

It is evident from Table 22.4 that the number of parameters in the CEA that require sensitivity analysis by user is smaller than that for most of the other evolutionary algorithms such as the GA (only parameters (1)–(3) require the sensitivity analysis). The sensitivity analysis of many parameters is implicitly done by the CEA during problem optimization. This makes the algorithm more flexible than others in solving a variety of optimization problems.

22.8 Termination Criteria

The termination criterion determines when to terminate the algorithm. Selecting a good termination criterion has an important role on the correct convergence of the algorithm. The number of iterations, the amount of improvement of the solution between consecutive iterations, and the run time are common termination criteria for the WCA.

22.9 Pseudocode of the CEA

```
Begin
  Input information of the algorithm and initial data
  Initialize the value of portions of selection and
    generation operators
  Generate initial population of possible solutions
  Evaluate fitness value for all solutions
  While (the termination criteria are not satisfied)
     Identify and memorize the best solutions (elites)
     For each selection operator
         Select solutions using the selected operator
           according to its proportion
         For each crossover operator
             Generate new solutions using the selected
               operator according to its proportion
         End For
         For each mutation operator
             Generate new solutions using the selected
               operator according to its proportion
         End For
     End For
     Set Population = new generated solutions + best
       selected solutions (elites)
     Evaluate fitness value for new solutions
     Update operator's proportions based on the amount
       of improvement of new generated solutions in
       comparison to the best solution
  End while
  Report the best solution
End
```

22.10 Conclusion

This chapter introduced a new meta-heuristic optimization algorithm called the CEA. A general understanding of the CEA was provided, followed by a description of the CEA's components. The CEA combines features of other algorithms and takes advantages of some of their best features.

References

Bozorg-Haddad, O., Adams, B. J., and Mariño, M. A. (2008). "Optimum rehabilitation strategy of water distribution systems using the HBMO algorithm." Journal of Water Supply: Research and Technology, AQUA, 57(5), 327–350.

Bozorg-Haddad, O., Afshar, A., and Mariño, M. A. (2006). "Honey-bees mating optimization (HBMO) algorithm: A new heuristic approach for water resources optimization." Water Resources Management, 20(5), 661–680.

Bozorg-Haddad, O., Afshar, A., and Mariño, M. A. (2011). "Multireservoir optimisation in discrete and continuous domains." Proceedings of the Institution of Civil Engineers: Water Management, 164(2), 57–72.

Bozorg-Haddad, O., Ghajarnia, N., Solgi, M., Loáiciga, H. A., and Mariño, M. A. (2016). "A DSS-based honeybee mating optimization (HBMO) algorithm for single- and multi-objective design of water distribution networks." In: Yang, X.-S., Bekdaş, G., and Nigdeli, S. M. (Eds.), Metaheuristics and optimization in civil engineering, Modeling and optimization in science and technologies, Vol. 7, Springer International Publishing, Cham, 199–233.

Bozorg-Haddad, O., Mirmomeni, M., and Mariño, M. A. (2010). "Optimal design of stepped spillways using the HBMO algorithm." Civil Engineering and Environmental Systems, 27(1), 81–94.

Dimou, C. and Koumousis, V. (2003). "Genetic algorithms in competitive environments." Journal of Computing in Civil Engineering, 17(3), 142–149.

Holland, J. H. (1975). "Adaptation in neural and artificial systems." 1st Edition, University of Michigan Press, Ann Arbor, MI.

Hormwichian, R., Kangrang, A., and Lamom, A. (2009). "A conditional genetic algorithm model for searching optimal reservoir rule curves." Journal of Applied Sciences, 9(19), 3575–3580.

Solgi, M., Bozorg-Haddad, O., and Loáiciga, H. A. (2016b). "The enhanced honey-bee mating optimization algorithm for water resources optimization." Water Resources Management, 31(3), 885–901.

Solgi, M., Bozorg-Haddad, O., Seifollahi-Aghmiuni, S., and Loáiciga, H. A. (2015). "Intermittent operation of water distribution networks considering equanimity and justice principles." Journal of Pipeline Systems Engineering and Practice, 6(4), 04015004.

Solgi, M., Bozorg-Haddad, O., Seifollahi-Aghmiuni, S., Ghasemi-Abiazani, P., and Loáiciga, H. A. (2016a). "Optimal operation of water distribution networks under water shortage considering water quality." Journal of Pipeline Systems Engineering and Practice, 7(3), 04016005.

Sonmez, R. and Uysal, F. (2014). "Backward-forward hybrid genetic algorithm for resource-constrained multiproject scheduling problem." Journal of Computing in Civil Engineering, 29(5), 04014072.

Wiley Series in Operations Research and Management Science

Operations Research and Management Science (ORMS) is a broad, interdisciplinary branch of applied mathematics concerned with improving the quality of decisions and processes and is a major component of the global modern movement towards the use of advanced analytics in industry and scientific research. The *Wiley Series in Operations Research and Management Science* features a broad collection of books that meet the varied needs of researchers, practitioners, policy makers, and students who use or need to improve their use of analytics. Reflecting the wide range of current research within the ORMS community, the Series encompasses application, methodology, and theory and provides coverage of both classical and cutting edge ORMS concepts and developments. Written by recognized international experts in the field, this collection is appropriate for students as well as professionals from private and public sectors including industry, government, and nonprofit organization who are interested in ORMS at a technical level. The Series is comprised of four sections: Analytics; Decision and Risk Analysis; Optimization Models; and Stochastic Models.

Advisory Editors • Analytics

Jennifer Bachner, Johns Hopkins University

Khim Yong Goh, National University of Singapore

Founding Series Editor

James J. Cochran, University of Alabama

Analytics

Yang and Lee • *Healthcare Analytics: From Data to Knowledge to Healthcare Improvement*

Meta-Heuristic and Evolutionary Algorithms for Engineering Optimization,
First Edition. Omid Bozorg-Haddad, Mohammad Solgi, and Hugo A. Loáiciga.

Attoh-Okine • *Big Data and Differential Privacy: Analysis Strategies for Railway Track Engineering*

Forthcoming Titles

Kong and Zhang • *Decision Analytics and Optimization in Disease Prevention and Treatment*

Decision and Risk Analysis

Barron • *Game Theory: An Introduction,* Second Edition

Brailsford, Churilov, and Dangerfield • *Discrete-Event Simulation and System Dynamics for Management Decision Making*

Johnson, Keisler, Solak, Turcotte, Bayram, and Drew • *Decision Science for Housing and Community Development: Localized and Evidence-Based Responses to Distressed Housing and Blighted Communities*

Mislick and Nussbaum • *Cost Estimation: Methods and Tools*

Forthcoming Titles

Aleman and Carter • *Healthcare Engineering*

Optimization Models

Ghiani, Laporte, and Musmanno • *Introduction to Logistics Systems Management,* Second Edition

Forthcoming Titles

Smith • *Learning Operations Research Through Puzzles and Games*

Tone • *Advances in DEA Theory and Applications: With Examples in Forecasting Models*

Stochastic Models

Ibe • Random Walk and Diffusion Processes

Forthcoming Titles

Donohue, Katok, and Leider • *The Handbook of Behavioral Operations*

Matis • *Applied Markov Based Modelling of Random Processes*

Index

In this document, page numbers referring to figures and those referring to tables are identified by putting in italics and in bold, respectively.

Meta-Heuristic and Evolutionary Algorithms for Engineering Optimization, First Edition. Omid Bozorg-Haddad, Mohammad Solgi, and Hugo A. Loáiciga.

d

e

h

i

k

l

m

o

p